W0018422

EXPRESSIVE ITERATION

A ground-breaking guide to thinking about how routine activities can be designed and innovated to develop narrative meaning and a sense of purpose.

Iteration is an integral part of daily routines, such as sleep-wake cycles, commuting, workouts, chores, or practising an instrument. While many iterations just monotonously repeat, others can lead to progression or evolution. With subtle variations among iterations, we can create meaning out of repetitive acts, forging narratives from them and thus making them meaningful to us. Chow draws on rhetoric, psychology, narratology, and design-thinking to show both in theory and in practice how we can innovate the design of mundane and routine activities to give them meaning and expression. He does so by examining Asian and European-originated examples, across a range of domains including visual arts, literature, digital art, video games, and mobile applications.

A must-read for designers and enthusiasts looking for ways to innovate across all domains and media and transform tedious repetitive activities into acts of intention.

Kenny K.N. Chow is Associate Professor in the School of Communication at Hong Kong Baptist University. His book, *Animation, Embodiment, and Digital Media: Human Experience of Technological Liveliness*, was selected as the Outstanding Book by the Korea Ministry of Culture, Sports and Tourism in 2021.

EXPRESSIVE ITERATION

Designing for Meaningful Routines

Kenny K.N. Chow

Routledge
Taylor & Francis Group

LONDON AND NEW YORK

Designed cover image: Katherine Luk

First published 2025
by Routledge
4 Park Square, Milton Park, Abingdon, Oxon OX14 4RN

and by Routledge
605 Third Avenue, New York, NY 10158

Routledge is an imprint of the Taylor & Francis Group, an informa business

British Library Cataloguing-in-Publication Data
A catalogue record for this book is available from the British Library

Library of Congress Cataloging-in-Publication Data
Names: Chow, Kenny K. N., 1971– author.
Title: Expressive iteration : designing for meaningful routines /
Kenny K.N. Chow.
Description: Abingdon, Oxon ; New York, NY : Routledge, 2025. | Includes bibliographical references and index.
Identifiers: LCCN 2024044371 (print) | LCCN 2024044372 (ebook) | ISBN 9781032489254 (hardback) | ISBN 9781032489247 (paperback) | ISBN 9781003391449 (ebook)
Subjects: LCSH: Human-computer interaction—Psychological aspects. | User interfaces (Computer systems)—Psychological aspects. | Technology—Social aspects. | Repetition (Philosophy)
Classification: LCC QA76.9.H85 C44114 2025 (print) | LCC QA76.9.H85 (ebook) | DDC 005.4/37—dc23/eng/20241018
LC record available at https://lccn.loc.gov/2024044371
LC ebook record available at https://lccn.loc.gov/2024044372

ISBN: 978-1-032-48925-4 (hbk)
ISBN: 978-1-032-48924-7 (pbk)
ISBN: 978-1-003-39144-9 (ebk)

DOI: 10.4324/9781003391449

Typeset in Times New Roman
by codeMantra

CONTENTS

ACKNOWLEDGMENTS

I would like to thank three people who have taken a brave first look at the initial draft of the book introduction and provided me with very useful feedback. Bruce Wan, a former colleague and friend of mine for over a decade, always lets me better relate my rhetoric to the designer's mind. Another former colleague who has shared thoughts with me is Jeffrey Ho. His analytical talent and academic integrity strengthen my belief that the book properly speaks to its target audience. Last but not least, Andy Cheung, an architect "slash" psychotherapist, also my longtime friend, never hesitates to express his views, which keep me ceaselessly contemplating the meaning of routines to different people from different walks of life. Without their stimulating comments, this book would never have been in this shape. I also very much appreciate the anonymous reviewers whose insights have greatly improved the framing of the book.

This book is a product of my multiple, parallel journeys of research interests and teaching development. I owe a special debt to many of my former students who participated in the exploratory design workshops with me and generated promising ideas that underpin many cases discussed in this book. Among others, I attempt to list the names of those whose works give rise to the cases. They are Adrian Stapfer, Anthony Tse, Careful Lam, Tom Lam, Chloe Leung, Qijia Chen, Leung Wing Yan, Ying Pak Chuen, Lam Wing Yan, Wang Yihan, Li Zhuoyang, Chan Cham Fung, Au Hiu Lui, Tam Cheuk Yi, Char Wai Ying, Ho Tsz Ching, Wagi Kulasumpankosol, Lana Dai, Megan Liao, Fan Sin Tat, Kuo Szuyu, Jaelynn Lam, Nia Pong, Kay Chan, Leung Sai Yin, Wong Sze Man, Yeung Wing Yan, Kelly Leung, and Esther Huang. I always learn a lot from my students. I also want to thank every expert participant who contributed to the evaluation study reported in the book. The study was supported by a grant from the Hong Kong Research Grant Council (Project No. HKBU 15608522).

Many illustrations in this book are images courtesy of a few talented digital artists and computer graphics animators, including Katherine Luk, Leung Wai Kit, Law Chi Wai, and Kwok Ka Yan. I thank them all. Special thanks are given to Shum Sheung Yin, who kindly contributed the calligraphy for the remake of the Chinese calendar charts presented in this book. Thanks to my copyeditor, John Lee, the text is clear and accessible. He helps me not only in writing but also in thinking. I would also like to thank Clarissa Lim and Khadijah Ebrahim in Routledge for their swift response and professional advice during the publishing process.

In the many months continuingly working on my desk for this book, I would never have made it through without my family, including my wife, Ida, and two cats. I thank them for being part of my meaningful routines. They are my reasons.

1

INTRODUCTION

In the Hollywood movie *Groundhog Day* (1993), the protagonist Phil Connors (played by the actor Bill Murray) is stuck in time and relives the same day over and over. He tries many different ways to break the time loop including killing himself, but everything will be reset the next morning and he will find himself waking up on the same bed on the same day again. Ironically, many people in self-quarantine during the COVID-19 pandemic seemed to have bizarre experiences comparable to Connors'. Waking up in the same room every morning, one had to repeat the same routine daily—having more or less the same breakfast, lunch, and dinner; turning on the same television and browsing the same channels; doing the same physical exercises; trying to finish reading the same book; showering in the same bathroom; sleeping on the same bed; and anticipating the same the next day. Fortunately, the quarantine period was usually finite and known in prior, unlike Connors' seemingly infinite time loop. And the classic novel *The Decameron* (1353) could shed some light on the matter. A self-quarantiner might as well set a daily storytelling goal, targeting to finish writing (or reading) one short story on a different subject matter every day. As the list of stories grows slowly, one could feel strongly that the end of the quarantine is approaching. After the quarantine, those stories constituted a collection like the famous novel. In the same vein, people photographed their every meal during the quarantine and shared them on social media like an art piece (Figure 1.1).

After the pandemic, we are free to live every day differently. Yet in reality, most of us still have to perform certain repetitive tasks or follow certain routines daily. After waking up, one brushes their teeth, has breakfast or coffee, gets dressed for work, commutes to work, and shops for a convenient and hopefully enjoyable meal for lunch. After work, one may still have certain chores that need to be done on a daily basis, such as walking their dog. These routines are basically "necessary" but

DOI: 10.4324/9781003391449-1

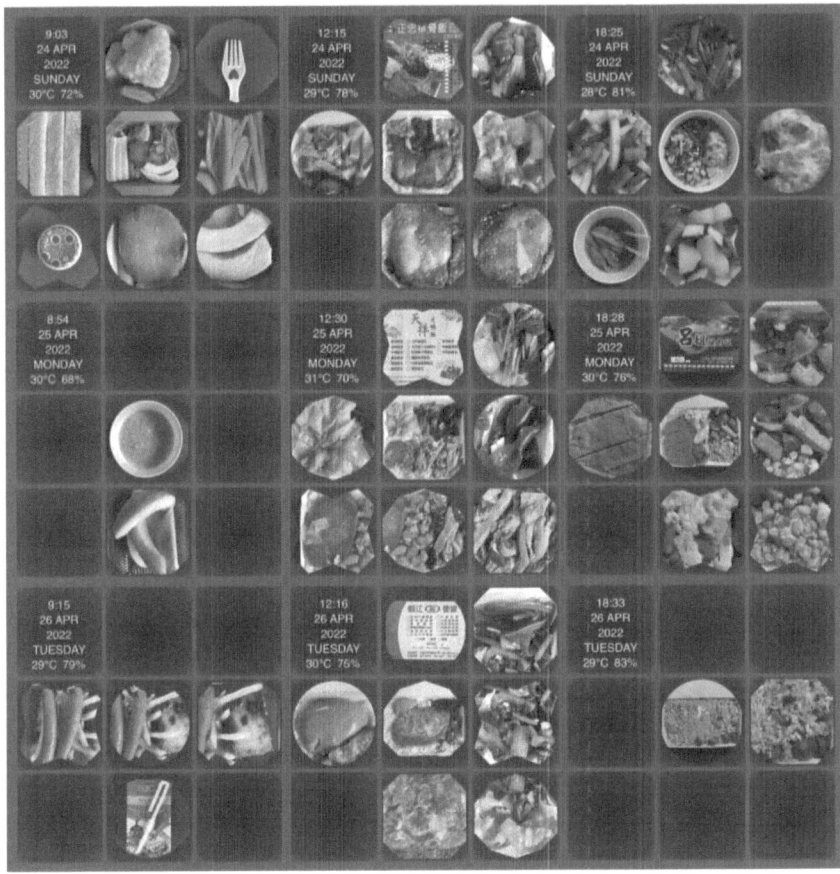

FIGURE 1.1 Part of the author's meals photographed during quarantine and arranged in a grid format.

also monotonous. Sometimes their lack of variation and boring nature make one question whether anyone of us can truly break away from the time loop in *Groundhog Day*. (This is exactly the question a student of mine raised.)

Other routines can be self-initiated, like a daily workout for physical fitness or reading a self-help book for mental wellness. Yet, such routines might not be engaging enough to keep one interested for extended period of time. The lack of noticeable improvement in physique or in mood may discourage someone from keeping them up. When one comes across temptations deemed more rewarding, they are more than ready to drop the routines.

Whether the routines are "necessary" or self-initiated, variation is crucial to keep someone motivated and sticking to them. As *Groundhog Day* and *The Decameron* hint at it, repetition can be imbued with variation. Through imagination and creativity, Connors explores various possible ways to relive the same day and finally finds an unexpected way out, whereas the sheltered people in *The Decameron* came up with various entertaining and stimulating stories to survive the plague in the fourteenth-century Europe. Hence, this book aims to explore whether and how we can make daily repetitive acts more engaging and even expressive. With subtle variations among repetitions, we can make meaning out of iteration.

Iteration for a Reason

Iteration, or repetition of processes, may sound boring. Yet, it can be powerful when coupled with variation. After the Big Bang took place about 14×10^9 years ago, atoms and molecules have undergone different chemical processes countless times, giving rise to vast galaxies, myriad stars, and innumerable planets, including the Earth. Since the birth of our blue planet and the solar system 4.5×10^9 years ago, natural phenomena like sunrise and sunset, the water cycle, and the cycle of seasons have repeated thousands of millions of times, nurturing various organisms. These organisms reproduce incessantly, with subtle and random variations of their genetic makeup, or so-called "mutations," which make different traits, breeds, and lineages possible. Some lineages were eliminated by natural selection, including the once-prevalent dinosaurs and many of our distant primate relatives in the African continent. Homo erectus left Africa some 1.5×10^6 years ago but finally went extinct. Following their predecessors, Homo sapiens tried many times to get out of Africa in various ways, but they repeatedly failed to survive in other continents. (Each migration might have taken thousands of years across generations.) Finally, the latest successful iteration happened 80×10^3 years ago and has brought about different human races across the globe today.

Every one of us is a product of the incomprehensible cycles of iteration before us. Yet, many of these cycles were never meant to produce any specific outcomes. The movements of molecules in natural phenomena were never directed by any biological agents for any purpose (possibly by celestial agents that we are too primitive to know). Reproduction is initiated by organisms, but mutation happens by accident, which may produce offspring that are better or worse in adapting to the environments. Natural selection is thus a complex process of replication with random variations that may or may not survive and reproduce. Only those adapting well to the environment can survive and reproduce, as if they are "selected" by super-intelligence. Migration can be driven by collective intelligence after numerous decisions are consciously made on an individual level. That said, when considered in retrospect, each step of the way (e.g., one crossing the valley for food) is by no means directly linked to the end results (e.g., one ending up moving to another continent). In short, each cycle of iteration plays a role in affecting later outcomes yet in an indirect and inscrutable way.

Although many iterative processes are uncontrollable (e.g., seasons) and even unpredictable (e.g., mutation), humans still manage to live with the situations through other manageable activities (e.g., farming different crops in different regions).

Human civilization is believed to have started about 10,000 years ago. Through repeating relevant processes with variations, people have constantly developed new techniques and principles in agriculture, medicine, construction, manufacturing, transportation, as well as the arts. Sometimes iteration can fix issues and turn failure into success,[1] like a baker repeatedly trying different ratios of flour, sugar, yeast, and water together with different fermentation times to achieve optimal rising despite the uncontrollable humidity and temperature during each season. This can be named "outcome-focused iteration" wherein the "successful" outcome of the last repetition is of utmost importance because it fixes the issues and makes things work. Other times, there is no specific issue to be fixed. People iterate an act because they want to explore new possibilities, like Claude Monet repeatedly painting haystacks at different times of day and in various seasons in order to study and depict the nuances of light and shadows, as well as the mood they deliver. In this kind of "process-focused iteration," or what I shall call "expressive iteration," every cycle of iteration and its intermediate outcome can show and tell a part of the story. Every painting in the *Haystacks* series by Monet is distinctive, and no single one can be considered more "successful" than the others. They altogether show the changes of light over time, which is the narrative of *Haystacks*.

It is said that Monet did not produce the *Haystacks* paintings one after another, but instead working on multiple canvases simultaneously,[2] because at certain time of day, such as dawn, the light changed so rapidly before he could finish a painting. Hence, he strived to capture the light for one painting and switched to another when the light had changed noticeably. With digital technology of today, artists can create an art piece much faster than Monet. The digital artist Beeple has been creating and posting one original digital image each day since 2007, in what he calls the *Everydays* project. After the first 5,000 days, Beeple compiled the set of images into an NFT (non-fungible token) sold at a record-setting price in the digital art market.[3] The themes of *Everydays* range from mixing icons or figures of pop culture in the beginning to making reference to the latest news in a futuristic science-fiction style later on. Being an experienced visual designer, Beeple is able to create a digital art image within a day. Unlike Monet, Beeple does not need to rush to capture the rapidly changing light with naked eyes but visualizes with the help of many software tools and references. The processes Monet repeated in producing *Haystacks* are specifically observing, documenting, and presenting the changes of light in nature. What Beeple repeats every day are more than mere production but also distribution (on social media) of his responses to the ever-changing stranger-than-fiction world.

Artists are used to iteration in creative practice, both in terms of techniques and subject matter. Monet and Beeple are not the exceptions, but more the norms. Many contemporary artists, such as those in serial art who produce series of works from iteration, adhere to a self-initiated plan with a set of rules (e.g., Sol LeWitt's *Incomplete Open Cubes*). Performing artists not only repeatedly practice and rehearse for better performance (i.e., outcome-focused iteration) but also tend to reinterpret, refashion, or revitalize a work of theatre or music for exploration (i.e., process-focused iteration). They iterate an act because they want to express something through the processes or outcomes. How about people other than practicing artists? Some people, unlike artists, might not always have something to express. Yet, they iterate an act regularly because they want to fix an issue (e.g., eating oatmeal every morning to lower blood cholesterol level) or make life better (e.g., doing exercise every day to build muscles). In other words, they iterate routines in daily life for positive outcomes.

Routine: Iteration in Daily Life

Routine is a set of actions that a person needs or wants to do on a regular basis. Many mundane routines are unavoidable, necessary, or favorable in daily life, such as getting up at the same time, brushing and flossing one's teeth, getting dressed properly, commuting to work or back home, staying focused on work and taking regular breaks, eating a healthy diet, drinking water regularly, washing hands after toilet, practicing yoga, walking the dog, taking a shower, and doing laundry or other chores. We iterate daily routines typically because we value the outcomes. For the desirable outcomes to emerge in the long run, such as physical fitness, mental wellness, or environmental sustainability, one is willing to endure the monotonous boredom and perform such acts repeatedly. Before long-term benefits become visible and tangible, immediate rewards, such as pleasure, are also necessary for one to enjoy once in a while and to continue the seemingly long journey.[4] Whether it is about immediate rewards or long-term benefits, positive outcomes constitute the reasons for practicing a routine.

Some of us may feel good about the immediate outcomes of some routines, like the minty taste in our mouth after brushing teeth with toothpaste. Yet, not every routine offers immediate rewards like the mint toothpaste. Very few people would enjoy the machine noise while vacuuming the floor. One might find the processes of a routine enjoyable, just like we enjoy our pet's company when walking a dog. But the novelty may wear out and the daily task becomes unattractive. One might start to procrastinate or be distracted by other options, such as playing a new video game instead of walking a dog. The fact that routines can be easily interfered can be explained by the science of habit, which informs that automatic repetition of actions requires rewards that are not only intrinsic and immediate but also unexpected or novel. This is due to the "reward prediction" in our brain wherein more dopamine is released for a quarter of a second when rewards are anticipated. The increased secretion of dopamine reinforces signal transmission along certain pathways (e.g., linking motor and sensory areas). If the anticipated reward occurs within one second, the dopamine level stays the same; otherwise (i.e., an error) the level drops in the next quarter of a second. In case of an unexpected reward, the level gets a boost for half a second to strengthen the pathways.[5] In short, unexpected rewards boost more dopamine secretion than expected ones. The bigger the unexpected reward, the more dopamine gets released, the stronger the sensorimotor pathways become, and the more automatic the relevant actions turn. Hence, while one might enjoy the ordinary processes or intrinsic outcomes of a routine, lack of new stimulus would still make one skip it easily in daily life.

The reward prediction phenomenon seems to resonate with the psychologist Erich Fromm's concept of "negative freedom" (1984). Humans are apparently "free from" instinctive and reflex action patterns found in animals, but still "tied to the world" when weighing and choosing among different possible courses of action (27). People choose to take certain actions because they believe in the possible outcomes. For instance, cycling to work is good for health and environment-friendly in the long run, but also requires getting up earlier every morning. The choices, though consciously made, are still dependent on many external factors in particular context. The dopamine reward prediction system in our brains then turns repeated actions with immediate, unexpected rewards into automatic habits, which no longer require our conscious thoughts for every later repetition. In principle, automaticity frees our minds from processing current mundane work for other high-level cognitive tasks[6] (e.g., thinking about where to meet a friend while mindlessly brushing teeth). Automaticity is efficient, but also mechanical and restrictive. It tempts us to give up control for now and let the past experiences take the lead. As Gretchen Rubin (2015) puts it, we decided not to decide (191). A proper choice consciously made in the past can be counter-intentional later. Yet, as automaticity effectively frees us from thinking about the current action, we become not "free to" think again on most occasions.

What Fromm (1984) calls "positive freedom" is that one is "free to" actively express the authentic self and experiences in emotion, thinking, and senses (222–3). Maybe few artists are lucky enough to enjoy this kind of freedom at times (223–4). They can focus on creative processes that express their genuine thoughts and feelings in tangible ways. Monet spent months on depicting what caught his eyes and mind most in the landscape; Beeple has been visualizing his responses to the world for more than a decade. Yet, many of us oftentimes still have to follow common practices in society, including daily routines, because the predictable outcomes give us certainty and security (e.g., commuting by metro to work on time to earn a living, doing laundry regularly for personal hygiene). While performing a routine, we may be free from (i.e., negative freedom) thinking about the repetitive acts by automaticity and seldom imagine any possible variation of the acts. Can we make a difference in each repetition? Can we leave a "fingerprint" on the processes? Can we add "color" reflecting our moods to each outcome? Can we make routine an expression of our authentic selves like how artists iterate to express? Although we are inevitably tied to worldly routine acts, we might explore the positive freedom to express by practicing them creatively.

One distinct artifact found in Beijing's imperial Chinese palace—Forbidden City—may suggest a way of making daily routines expressive. It is a rectangular board about the size of a sheet of A2 paper mounted on a wall. The board features nine different Chinese characters arranged in a three-by-three matrix

FIGURE 1.2   A remake of a Chinese poetry-based winter calendar chart originating in imperial China. Calligraphy by Shum Sheung Yin.

(see Figure 1.2). Following the Chinese vertical writing tradition, they look like a Chinese poem, which can be literally read as "in front of the pavilion, weeping willows, take care, and waiting for the breeze of spring." It is actually a winter calendar chart. The winter lasted for 81 days in the Chinese calendar. Each of the nine characters consists of nine strokes, adding up to 81 strokes. Imagine on a winter solstice more than 100 years ago, someone in the palace started to write one stroke each day. By the ninth day, the first character "pavilion" appeared at the top-right corner of the matrix. This pictographic Chinese character acted as an intermediate virtual reward for the person who counted the days. Other vivid characters like "willow" (the top-middle), "spring" (the middle-left), and "breeze" (the bottom-left) followed accordingly. Every new character evoked a new image that soothes one's soul along the journey. When a hopeful, vivid poem was completed after 81 days, one had survived the harsh winter. Similar poems existed as folk art as well.[7] This distinctive Chinese poetry-based winter calendar chart intriguingly turned the ordinary, day-to-day act of marking into a process of revealing a poetic mini-story: on a winter day, we parted with someone amid weeping willows in front of a pavilion. The story gave people a reason to iterate the routine every winter day. One wanted to complete the chart and be rewarded by the sight of a full poem.

The Chinese poetry-based winter calendar chart enables anyone to iterate a mundane act in a way like a calligraphic artist or poet does. Instead of showing the completed chart only in a private place like the palace, one may imagine presenting it publicly, like what most artists commonly do. Monet might rely on an art dealer, while Beeple needs only Internet access. After the advent of digital technology and social media, we are never more than "one click" or "one tap" away from disseminating information to almost everyone around the globe. A digital platform can be designed to engage people in a daily routine and allow them to instantly share their performance online. An excellent example is the web-based word puzzle game Wordle, which became a sudden hit in 2021 and then was acquired by *The New York Times* in early 2022.[8] The game greets players with a five-letter word puzzle every day and players have to guess the word in six tries. Every attempt is given feedback with colored hints wherein green means correct letter in correct spot, yellow means correct letter in wrong spot, and gray means "forget it." They result in a colored grid, making it possible to share one's performance that day on social media, without revealing the answer. A player is motivated to take the daily puzzle challenge and show off their performance. There may be "fear of missing out" the word of today, especially when one sees their friends' or celebrities' solving the puzzle on social media. Players perform this daily routine, because they not only enjoy the guessing process but also want to record and show off their daily achievements. In this regard, Wordle players share some common threads with Beeple. The artist takes a self-initiated challenge, be it technical or conceptual, to create and post an image every day, while players of Wordle take a given challenge to guess and post the result once a day.

Iterate to Express

Monet's *Haystacks*, Beeple's *Everydays*, the Chinese poetry-based calendar chart, and Wordle are exemplars of what I call "expressive iteration." There are some commonalities among them in their processes.

First, someone performs a procedure on a regular basis, mostly every day, with variation. Monet painted the same subject at different times of day for many consecutive days. The paintings showed similar forms but varied in color. Beeple created one original digital image every day and posted it online. The images varied in theme. A person living in Forbidden City wrote one stroke of a character on the winter calendar chart every day. A player of Wordle tries five letters in every guess to solve the puzzle and shares the result online every day. The results follow the same color-coding scheme but show different patterns.

Second, someone produces a series of outcomes resulted from iteration, which elicit imaginative images on one's mind. The *Haystacks* series projects a tranquil field wherein atmosphere keeps changing. Images of *Everydays* depict different aspects of an eccentric, futuristic, sometimes dystopian, world. The different Chinese characters written on the winter calendar chart form a poem describing objects, actions, setting, and emotions. Color-coded results of Wordle outline how close each guess is and with how many attempts the puzzle is solved.

Lastly, someone directly or indirectly presents the serial outcomes as a tangible narrative to satisfy oneself or appeal to others. Through art dealers, Monet showed his *Haystacks* series as a visual diary of his meticulous study of light. Beeple impressed others by putting together all the images that he created in 5,000 days into a life-echoing journal. The Chinese poetry-based calendar chart was mounted on the wall for any passers-by to appreciate and contemplate the poetic scene. The results of playing Wordle posted online show in terms of color patterns a player's course of performance over a period of time.

The above cases demonstrate how individuals iterate a process and express themselves by showing a series of outcomes. Expressive iteration is thus a repetitive procedure that, in a lasting manner, speaks on behalf of the person who performs it. The process gives rise to serial outcomes as a tangible narrative. Variations between successive cycles of iteration constitute a "pattern" in the narrative, such as the change in light direction, angle and color in *Haystacks*, or each nine-stroke character in the Chinese winter calendar chart. By observing the pattern, one can see progress made in preceding cycles and imagine possible future actions and outcomes. This is a satisfying experience as it allows room for imagination. It also gives people a new reason for repeating the procedure—to contribute to the continuous development of the narrative. Monet, when reviewing and appreciating those partly painted late autumn haystacks, might have a creative impulse to paint the early winter counterparts. Looking at the seven Chinese characters already written on the chart that evoke images of "pavilion" and "willows," a person would look forward to "the spring breeze," the final two characters. One has all the reasons for writing on the chart in following days to complete and read the hopeful poem. Hence, a person iterates not only to achieve original goals like counting the days to spring but also to express thoughts and feelings in tangible forms such as calligraphy and poetry. These varied, successive tangible records can serve as evidence of making progress, satisfying one's basic psychological need for competence, which is recognized as one of the internal driving forces, called "intrinsic motivation," regardless of external influence from environments.[9]

Meanwhile, the tangible narrative provides possibility of appealing to other people. This kind of record is not just relevant to artists like Monet or Beeple. Some players of Wordle intentionally produce visually distinct results, such as the same color pattern across consecutive unsuccessful attempts, which makes others wonder how many incorrect words with similarly spelling exist. One might want to guess what the correct word is. Those who have dropped the game recently might pick it up again out of curiosity. In this sense, a narrative of distinct, novel records can become highly persuasive. Posting interesting Wordle results may drive other potential players to creatively appropriate the game by inputting different words, not to solve the puzzle, but to target at creating results that paint a particular color pattern. For instance, making all squares in only gray (i.e., not even one letter is correct) can be as challenging as getting an all-green row (i.e., the correct word). There is a variant web application of Wordle[10] where the goal is not to solve the puzzle (the correct word is shown upfront) but to generate a specific color pattern with words that contain the correct and incorrect letters as required. One could collect different color patterns after days and show off the successful "incorrect" attempts made. All in all, people iterate a process and share the outcomes, which encourages others to join the iteration. After repeated practice, some may appropriate the process and generate distinct, appealing outcomes. Lately, the need for novelty has been proposed and tested as a new candidate of basic psychological needs alongside competence, autonomy, and relatedness as informed by Self-Determination Theory.[11] Expressive iteration can satisfy all these fundamental needs, including novelty (everyone can generate distinct outcomes), competence (one can demonstrate making of progress through the serial outcomes), autonomy (one can decide when to continue and how to appropriate the processes), and relatedness (one can share the outcomes with others as self-expression or invitations), enhancing intrinsic motivation.

New Reasons for Routines

If expressive iteration is applied in daily life, a mundane routine can be more engaging. A person typically needs to brush their teeth, commute to work and back home, eat and drink healthily, perform house chores, take a shower, and go to bed every day. Some of us may want to practice physical and mental exercises regularly, such as working out, stretching, or meditating. Some others may need to check in with their vital signs of mind-body wellbeing, such as weight, blood glucose level, peak expiratory flow, and even emotions. We know the primary reasons for sticking to different routines, including benefits in hygiene, health, life, or personal growth. Yet, it takes major self-discipline and perseverance to maintain them. First, benefits of many routines are long-term and it takes time to see any noticeable changes. That makes procrastination or suspension tempting. Skipping a workout or indulging in junk food for a few days might not immediately bring extra weight or undermine physical health. Ceasing to monitor blood glucose levels may even allow oneself to falsely assume that the level stays "normal" as one does not feel unwell. Second, some routines require more conscious effort, mental or physical, and can hardly become automatic like habits. Very few people would "automatically" do laundry, take a finger-prick blood glucose test, or input feelings or thoughts into a mood-tracking application. Parents might know how challenging it is to get kids to brush their teeth every day and night, not to mention getting them to take their daily vitamins. Sometimes, people need immediate rewards as secondary reasons for performing and maintaining healthy routines. There have been attempts of gamifying certain activities to encourage people to build or alter habits. After a task is performed, rewards like points are granted as new reasons for sticking to a routine. Such points can be accumulated to escalate the person to the next levels or exchanged for badges indicating progress or achievements. Although these basic game design elements are widely adopted in behavioral interventions,[12] studies show that they are too extrinsic and may not produce long-term effects.[13] Meanwhile, some research suggests using stories[14] or metaphors[15] to better facilitate intrinsic motivation.

On the other hand, expressive iteration offers people alternative new reasons to stick to a routine by turning it into an imaginative and satisfying story. As mentioned, expressive iteration satisfies basic psychological needs with its iterative processes and serial outcomes, enhancing intrinsic motivation. Further to that, the serial, tangible outcomes can show and tell stories. Some tangible outcomes can be figurative in their mode of representation, like Monet's *Haystacks* paintings; others can be iconic, like the Chinese characters in the winter calendar chart. Although being varied in the representational continuum, they can evoke vivid images on one's mind. With a series of mental images, one can easily and frequently imagine being in an event. This is sometimes called "mental simulation" by psychologists and has proved to render imagined events subjectively more likely.[16] When a sequence of these image-laden events constitute a story, and a person is so absorbed in it, "narrative transportation" may further take place.[17] One not just temporarily accepts the fictional world but also easily takes the underlying beliefs with them to the real world.[18] In other words, expressive iteration evokes images and stories, which can facilitate the linkage of a virtual item (e.g., an image of a pavilion) with a routine act (e.g., marking on a calendar). The former can feel like a part of the latter, that is, what Wendy Wood (2019) calls "intrinsic reward" that can be effective in motivating a behavior (118–9).

While the use of a story with images is powerful in affecting people, it can be distracting too.[19] At different cycles of iteration, a person can be absorbed in the story, but one should also be able to see its correspondence to the routine. If the story is not analogous to the routine, one cannot connect different stages of the story with the progress made in the routine. Hence, an appropriate story should involve actions comparable to routine acts. That makes performing a routine feel like acting out the story. Outcomes in the story can then be felt like a "natural consequence" of the routine acts.[20] Furthermore, the successive outcomes (e.g., the images of "willows," "spring breeze") should also be aligned with the long-term vision[21] (e.g., looking forward to the next spring), that is, the main reason for the routine. Long-term benefits of daily routines, such as physical fitness or mental wellness, can be abstract. For instance, it is hard to specify the goal of physical fitness without using a set of health-related data, and the progress one has made toward this goal cannot be easily defined. Conversely, the outcomes at different stages of the story can be more specific. Metaphorical projections from the fictional outcomes to the real-life long-term benefits can effectively help one understand the progress. In fact, many of our basic yet abstract concepts are shaped by metaphorical projections of recurrent concrete experiences, technically termed "conceptual or embodied metaphor."[22] For example, we say "high quality" to describe something good, because we often climb up (a bodily experience across space and time) to higher places for something better (an abstract idea) such as more spectacular views or quieter environments. In summary, an appropriate analogous story draws on similar actions in a routine, and specific outcomes of successive events in the story can correspond to an abstract progress toward the goal of performing the routine. If the fictional outcomes and the real-life long-term vision are coherent (without logical conflicts), the story is promising for expressive iteration.

Let us consider a mini-story of writing a Chinese poem wherein the main character is engaged in actions (e.g., using an ink brush to absorb ink, writing a stroke on paper), which feel like the routine act of blood glucose monitoring (e.g., using a paper strip to absorb a blood sample from the pricked finger, placing it in the meter) but in a different context (Figure 1.3). While a person regularly performs the routine (e.g., taking a blood glucose test) and repeatedly see the story outcomes (e.g., one more stroke toward finishing a Chinese pictograph in the poem), one would find it easy to imagine that blood glucose monitoring yields the poem. Frequency and ease of imagining an event, which can be supported by mixed-reality technologies of today (e.g., a wall-mounted digital display simulating a board), renders the

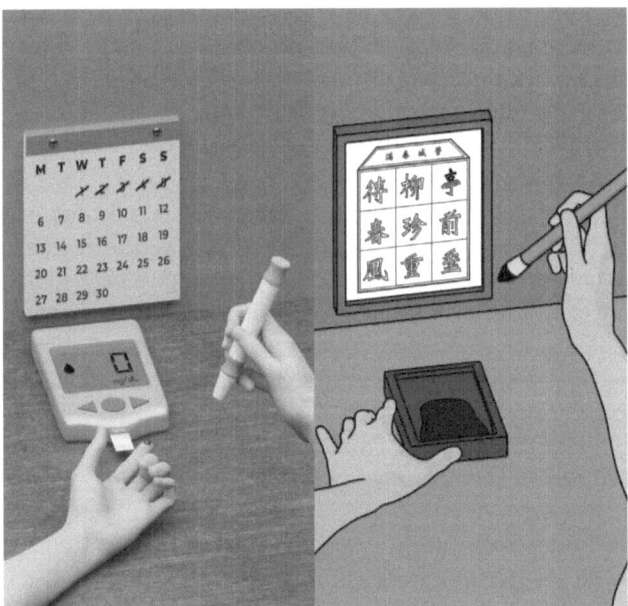

FIGURE 1.3   The act of using a paper strip to absorb blood and transferring to the meter in blood glucose monitoring (left) is comparable to the act of using an inkbrush to absorb ink and writing on paper when practicing Chinese calligraphy (right).

imagined event subjectively more likely.[23] The person feels that one can make the story unfold by performing the routine, which becomes the new reason for doing it. The written Chinese poem can be seen as a vivid virtual reward of blood glucose monitoring, yet the imaginative and satisfying story is the actual key. Iconic badges, if thoughtfully designed with a metaphorical story, can thus be useful. For example, the activity tracker Fitbit rewards users for walking or climbing stairs with badges of different shoes from the casual to the professional and even whimsical, as well as different destinations from iconic tourist spots to fantasy adventures.[24] The shoes and destinations seem to suggest a background story of journeys on foot. The more steps one has walked or stairs one has climbed, the more varied journeys the story presents. The story development offers a new reason why one climbs stairs to work every day.

Connecting a routine with an imaginative story can create new reasons for sticking to it. If a person can regularly and easily imagine the change they can make in the story world while performing the routine, one feels competent (a recognized basic psychological need) to make progress, albeit virtually in the story. Furthermore, the person may develop creative ways of manipulating outcomes in the story, acquiring a sense of autonomy (another basic psychological need). One thus makes a new meaning of the routine. Toward this end, the routine should have actions and objects reminiscent of the imaginative story, probably with the creative use of technology. For instance, blood glucose monitoring requires a person to regularly prick a finger, squeeze a drop of blood onto a test strip, and place it in the meter. Touching the blood drop on a test strip can evoke a person's memory of using an inkbrush to pick up the ink. Similarly, inserting the test strip into a blood glucose meter is reminiscent of writing a stroke on the calendar board. Holding the finger-prick device is like holding the inkbrush; touching the blood drop on a test strip is like picking up the ink. With accessible technologies such as contactless sensors, wireless connection protocols, and digital displays, we can create a novel mixed-reality experience that one stroke of a character appears whenever a finger-prick blood test is done. Completing and appreciating a vivid and hopeful poem becomes a new reason for doing the test regularly. The routine is mundane, and the imaginative story can be a slice of life. They are seemingly unrelated. Yet, uncovering their subtle connections and highlighting them with technologies generate a sense of novelty. This is a kind of innovation that not just improves what we are regularly doing but instead creates new reasons for our routines and makes new meaning of them.

Innovation, with its Latin origin, means "make new." An innovation can be defined as any novel product, service, or method that is deemed better than current practice.[25] Some marketing research scholars attempt to differentiate technological from design innovations. Technological innovation refers to new functionalities, concerns performance, and aims to prevent, while design innovation refers to new appearance, concerns social acceptance, and aims to promote.[26] In the design disciplines, innovation is commonly framed as making changes in a context. In *Vision in Design* (2011), Paul Hekkert and Matthijs van Dijk propose a framework that guides the design processes. According to them, design innovation is not only about changing the product appearance but also product interactions through analyzing past or current scenarios of use and synthesizing future ones. Ezio Manzini, on the other hand, focuses on changes in social context in his work *Design, When Everybody Designs* (2015). It describes how designers can support projects of collective individuals in making social changes. Roberto Verganti suggests a perspective of innovation that emphasizes the change of meaning in *Overcrowded* (2016), and this angle is more relevant to expressive iteration than others. He introduces two kinds of innovation. Innovation of solutions brings about improvements in performing a task or taking an action. Innovation of meaning offers novel interpretations that provide new reasons for taking an action. The former changes current practice, while the latter changes current meaning. By "meaning," Verganti refers to the purpose people try to achieve, that is, why they do certain things (32). For instance, Swatch, the watchmaker, changes the meaning (i.e., purpose) of wearing a watch from telling time to matching outfit. A fan wears a Swatch, because they want to match it with their cap or makeup. Nespresso, the coffee machine, updates the meaning of machine-prepared coffee from convenience and quality to flavor variety. People enjoy Nespresso, because they can pick different flavors at different times. iPhone expands the meaning of taking pictures from keeping memories to communication and expression. People use smartphone cameras to snap their breakfast, because they want to show and tell others what they are enjoying. While the smartphone camera is an innovation of meaning, many photo- or video-based instant messaging social networking apps, such as Instagram, Snapchat, and Tik-Tok, seem to offer varied innovative solutions providing different user experiences yet delivering the same new meaning. In this sense, innovation of meaning casts light in a new direction leading to a change of meaning, and then innovation of solutions suggests new ideas along that direction.

The above innovation frameworks cover the changes in product interactions brought along by designers, the changes in social contexts effected by designers and non-design communities, and the changes of meaning produced by businesses. Expressive iteration, as mentioned, through matching a daily routine with an imaginative story gives people new reasons for sticking to it and allows one to make new meaning of it. In other words, expressive iteration demonstrates the innovation of meaning by individuals in their daily routines.

Searching for Meaningful Stories

Expressive iteration provides new meanings to a routine by matching it with an imaginative story. Along this trajectory, searching for an appropriate story becomes the next objective. Following one of the common design thinking processes, the search consists of two parts, namely, the divergence and then the convergence.[27] First, one can start with imagining a list of possible scenarios that are analogous to the routine. Actions, alongside the objects and sensations, involved in the routine

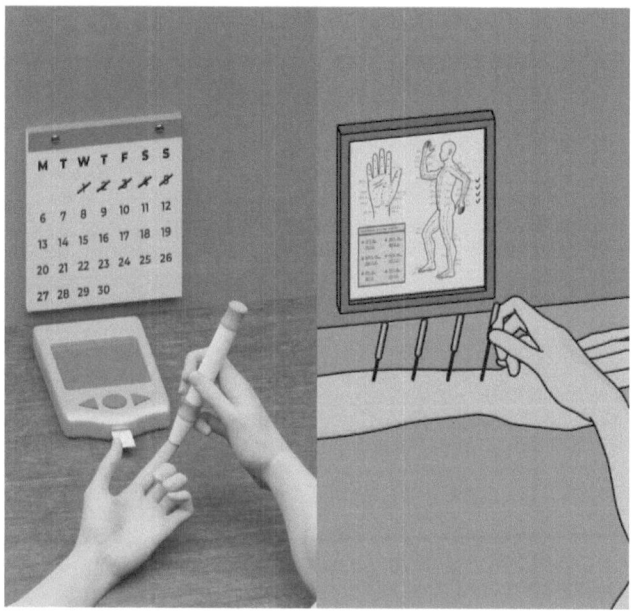

FIGURE 1.4   The association between blood glucose monitoring (left) and acupuncture
(right) in terms of the prick, the pain, and the needle.

are good jumping-off points. So are their iterative patterns. Reconsider the case of
blood glucose monitoring. The act of pricking, the pain, the needle, the blood drop,
the test strip, the blood glucose meter, and the reading may inspire different stories
in different genres. For example, the prick, the pain, and the needle are reminiscent
of acupuncture (Figure 1.4). Together with the meter reading, one might recall
tattooing (Figure 1.5). The needle and the prick may remind someone of embroidery (Figure 1.6). The blood drop and the test strip may also evoke the image of
calligraphy. The act of placing the test strip in the meter is not unlike feeding a
time card into a punch clock (Figure 1.7). Each of these mental associations has a
comparable iterative action pattern with the blood glucose test. Acupuncture, tattooing, and embroidery all involve repetitive acts of inserting a needle similar to
that of blood glucose monitoring, although the time intervals between repetitions
are different. Calligraphy requires repeated practice. Punch clocks and time cards
are used for tracking employee daily attendance in a workplace. These associations
thus can be developed into narratives with recurring events, such as a protagonist
regularly having new tattoos on the body for memorable life encounters (alluding
to Christopher Nolan's *Memento*), or writing new poems in response to changing
weather and mood. They may vary in genre from a slice of life, romance, adventure, thriller, to even science fiction. Searching for potential story ideas requires
awareness of and attention to common experiences with objects or environments.
On what occasion would one put a needle through? What would one use to pick

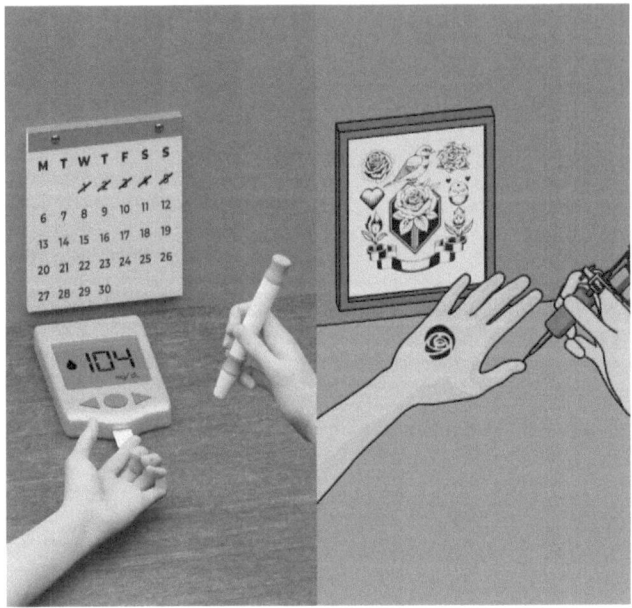

FIGURE 1.5   The association between the meter reading in blood glucose monitoring (left) and the graphic marked on skin in tattooing (right).

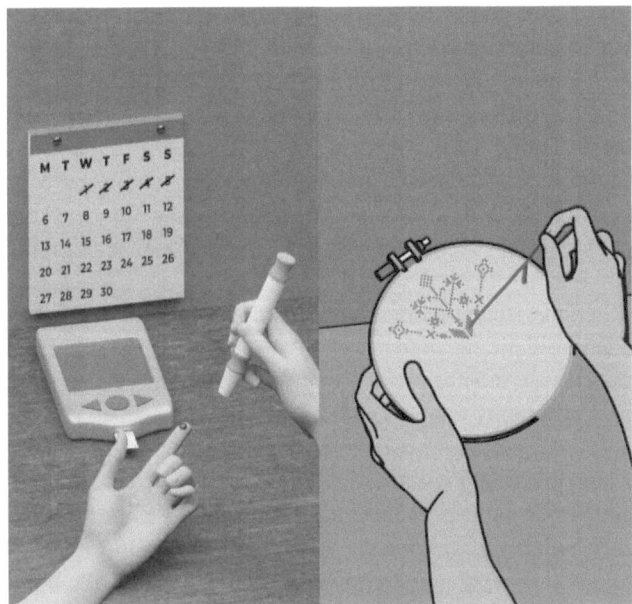

FIGURE 1.6   The association between blood glucose monitoring (left) and embroidery (right) in terms of the needle and the prick.

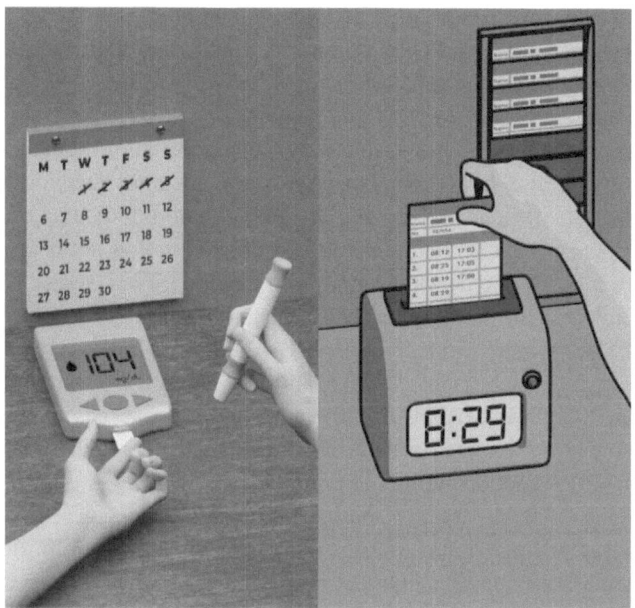

FIGURE 1.7   The association between placing the test strip into the meter for a reading (left) and feeding a time card into a punch clock for an attendance record (right).

up a small drop of liquid? One needs to "browse" through the mental inventory to look for comparable experiences including what we see, hear, and touch, or how we act and respond. For more possible associations, one has to accumulate contents for the library.

In *The Seventh Sense* (2015), William Duggan emphasizes collecting examples from history to build the inventory, meanwhile being mindful (e.g., stretching mental time) in daily life. On the one hand, there are many insightful cases of expressive iteration across different media domains, cultures, and periods. Apart from Monet's *Haystacks*, Beeple's *Everydays*, the Chinese poetry-based calendar chart, and Wordle, other examples can be found in visual arts, literature, digital art, video games, and mobile applications. They may originate from different places of the globe at different times. We need a framework to organize this diverse set of conceptual resources for effective access to any particular case. Introducing this framework will be one goal of this book. On the other hand, we need to be mindful of our bodily experiences in daily life, which also demands a way of dissecting experience into sensory perception (what we see, hear, touch, etc.), motor action (how we act and respond), as well as the sense of spatial relations (e.g., container, verticality, proximity) and temporal patterns (e.g., episodic, continuous). When performing a routine, a person can pay more attention to the above aspects of the bodily experience (i.e., stretching mental time on the routine). With mentally accessible references of expressive iteration in the library and focused attention to varied aspects of bodily experiences, one can readily and intuitively associate the routine with a set of analogous stories.

The second part of the search is a convergence process. Apart from comparable iterative action patterns, successive outcomes of a good story should also be in line with a person's long-term vision. The immediate outcomes function like intrinsic rewards, or treats, for one to keep up. Yet, they should not distract one from their own vision. Acquiring different rewards along the journey is not only encouraging but also reminds one of the original goals of the routine. For instance, a person who needs to regularly self-monitor their blood glucose level is likely to suffer from diabetes or similar chronic health conditions. Symptoms include having blurry vision and numb hands, which widely disrupt daily life activities. Maintaining a normal blood glucose level would allow one to pursue hobbies that require eye-hand coordination, such as calligraphy or embroidery. With the knowledge frame of diabetes symptoms invoked, the stories of practicing calligraphy or embroidery can be good analogous matches for this routine. When seeing the successive outcomes, either different calligraphed characters showing part of a poem or an unfinished stitchwork of a flower pattern, one can project the tangible work-in-progress onto a long-term vision of maintaining good eyesight and hand dexterity for engaging in daily activity and particularly enjoying the hobby. The invoked knowledge frame facilitates aligning the successive outcomes with the vision. Finishing and collecting poems or embroideries over time implies maintaining health over the same period of time. The calligraphed characters or embroidered stitch patterns not only represent the progress one has made toward the long-term goal but also function as intrinsic virtual rewards with a meaningful background story, especially if the hobby is of one's interest. They provide a new reason for self-monitoring blood glucose levels and embody the new meaning of the routine.

How This Book Is Organized

The purpose of this book is to suggest and demonstrate innovative ways of transforming mundane routines into serial, tangible narratives, which are imaginative, self-satisfying, and appealing to others. It aims to help readers connect a specific routine with an analogous story and find their new reasons for performing it. The book consists of three parts, corresponding to theory, case study, and design.

In Part I, the author theoretically characterizes expressive iteration and explains why it works. Expressive iteration persuades and motivates through iterative action processes and serial, tangible outcomes. Expressive iteration elicits imagination through showing and telling stories. Expressive iteration connects life routines and fictional stories through mapping (pattern matching). The proposition of expressive iteration in daily life is grounded in theories regarding rhetoric, persuasion, motivation, imagination, narrative, fiction, and analogy, which will be articulated in three chapters. Part I concludes that expressive iteration is characterized by iterative processes and serial outcomes, which present stories in vivid images, with projection of patterns between routines and stories.

Part II covers interpretations and analyses of a corpus of exemplary cases of expressive iteration from different domains, cultures, and periods. They include literary works like folktales and novels across cultures, Japanese woodblock prints and board games, Chinese calendar charts, serial art, film or television series, video games, and digital applications. Their origins vary in terms of place and time. These exemplars typically feature a pattern of iteration in their narratives, which can be projected onto a real-life event that each of them hints at. They are clustered into three levels of iteration, which will be described accordingly in three chapters. Studying their patterns will inform design considerations for creating and composing stories with iterative patterns that match the routine we want to make a new meaning of.

In Part III, we explore ways to give new meanings to routines using expressive iteration. The processes include collecting and comparing potential stories for matching a specific routine. I have developed initial design guidelines and conducted a series of design workshops on related topics, generating a collection of design prototypes since 2017. These prototypes demonstrate interesting ways of eliciting people's imagination and supporting the construction of their own relevant stories in fictional and virtual worlds. The final two chapters summarize the evaluation of the collection and map out different routines and promising story ideas, which are made possible by emergent technologies like mixed reality (MR) and generative artificial intelligence (AI). This final part also looks forward to our future stories.

PART I

Theory

Iteration literally means repetition of a process or an utterance. One may repeat a process to fine-tune the result or to double-check the result for accuracy. For example, a dancer rehearses a performance; an economist calculates the derivative of a utility function to second order in order to see the trend; or a programmer writes a "for-loop" to generate a fractal pattern with details in progressively smaller scales. On the other hand, someone may repeat an utterance to emphasize or clarify a message, so as to persuade the audience. For example, a parent asks their child to go to bed repeatedly; a suspect insists that they were not at a crime scene; a student talks incessantly to comfort themselves after getting their grade in an exam.

Someone may also regularly perform an act to demonstrate potential or explore possibilities. For example, a person posts their running path on social media every day. This action iteration shares commonality with utterance iteration, but there are nuances in the reasons behind. While iteration of an utterance sends the same message again and again to persuade others, iteration of an action delivers a series of outcomes to demonstrate. When the serial outcomes are presented in tangible forms, they function like intrinsic rewards for making progress or variations that satisfy one's innate psychological needs for competence and autonomy. They are visible social proof that appeals to others about possibilities. In short, this kind of action iteration persuades or motivates others with serial, tangible outcomes (Figure I.1). Chapter 2 focuses on this part of expressive iteration from the perspective of rhetoric, design, and the psychology of behavior.

The serial outcomes in tangible forms elicit mental images. Seeing a post on social media that displays a running path, a person can visualize how someone might run along a promenade, across a road, through an alley, over a footbridge, or by the side of a lake. When a series of mental images constitute a story, one can further be absorbed into it. For instance, if photos are attached along the running

DOI: 10.4324/9781003391449-2

FIGURE I.1 The persuasive process of expressive iteration to be discussed in Chapter 2.

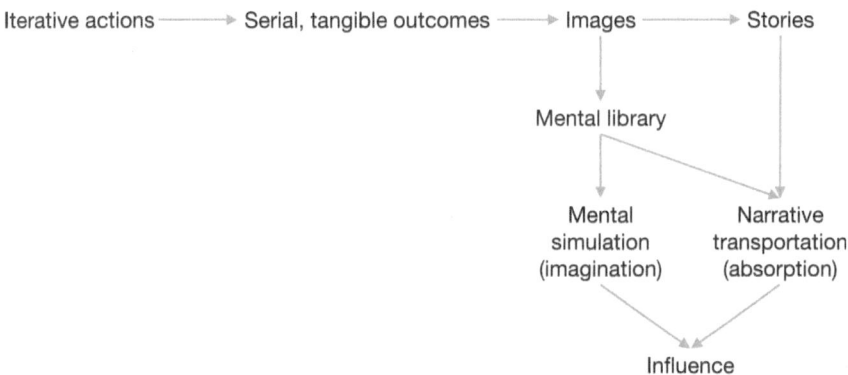

FIGURE I.2 The imaginative process of expressive iteration to be discussed in Chapter 3.

path, such as showing a sunset by the lake, rain over the footbridge, or a night view along the promenade, one can imagine someone running on the same path at different times of day, or even in different seasons. One might mentally transport oneself onto a story world wherein the cityscape changes over time, echoing what Monet observed in the wheat field over 100 years ago. These mental processes prove to affect one's beliefs and intentions, even in the case of fictional stories. Hence, when the serial, tangible outcomes of iterative action processes show and tell stories, they not only persuade but also influence (Figure I.2). Chapter 3 looks at this part of expressive iteration, and the implications.

Fictional stories with vivid images can absorb and affect the audience, but they might also distract one from their vision in real life. A story emerging from serial, tangible outcomes of iterative action processes should be analogous to a target routine. Analogy between a fictional story and the real-life routine builds on their similarities or correspondences in actions, objects, or outcomes across

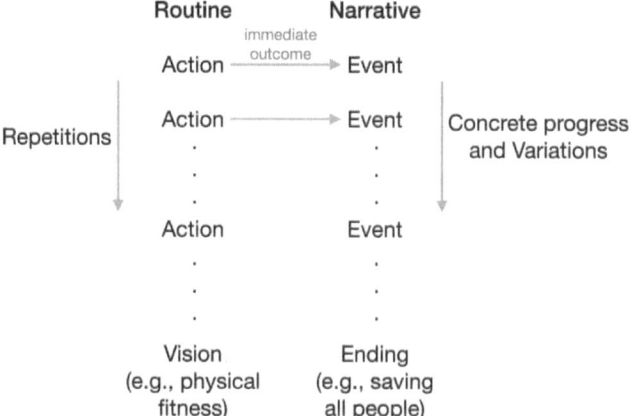

FIGURE I.3   The analogical process of expressive iteration to be discussed in Chapter 4.

a series of episodes. It should be easy for one to map a stage or event in the story onto their real-life experience, and the story ending should align with one's vision. If a daily running routine is connected to a fictional story of running from life-threatening zombies and collecting life-saving potions, as presented by the fitness mobile app Zombies, Run!, one should be able to map the amount or power of the collected potions onto the cumulative running distance or the best running speed. And saving all the people in the city from zombies is arguably in line with the reason behind the running exercise, that is, physical fitness. Serial, tangible outcomes from fictional events (i.e., virtual rewards) are aptly mapped onto the real-life cycles of the routine. Such analogical mapping allows interpretation of the virtual rewards acquired in the story as new reasons for sticking to the routine (Figure I.3). Chapter 4 addresses this part of expressive iteration, and its implications.

2

PERSUASION THROUGH PROCESSES AND OUTCOMES

Iteration of an action, like iteration of an utterance, can be used for persuasion on certain occasions. When someone hikes every weekend and posts on social media the scenic pictures they take along a trail together with a route map and an elevation profile, he or she not only intends to keep a record of their activities but also hopes to share with others the enjoyable journeys. Browsing the pictorial records over weeks, months, or even years in a digital album is like virtually revisiting the trails, and the tiling thumbnails give one a sense of ownership and satisfaction. The beautiful views and the level of difficulty in terms of distance and elevation, on the other hand, may invite or inspire others to do something similar. For instance, an expat wanting to explore the nature near the city, or parents looking for weekend plans for their family may follow suit. In short, the serial, tangible outcomes of this action iteration cause one to believe about one's own ability. They may also pique others' interest in the activity. This action iteration is persuasive to both the person performing the routine as well as others.

Believe It or Not

This kind of action iteration embodies persuasion in action processes and tangible outcomes. Persuasion, in its original sense, is effectuated by language communication. We speak or write to present arguments in order to convince someone or to make someone believe in something. Other modalities, like facial expression, gesture, handwriting quality, punctuation, or typography, sometimes generate additional persuasive effects such as emotion or emphasis. Dating back to the period of ancient Greece, influencing others' thoughts was the key purpose of public speaking. Philosophers of their times, including Aristotle, introduced the term "rhetoric" as the art or practice of persuasion. Their ideas, largely summarized as "classical

DOI: 10.4324/9781003391449-3

rhetoric" today, are characterized by three aspects, namely, logos (reason), ethos (credibility), and pathos (emotion).[1] Logos refers to the internal consistency and clarity of the argument, and how it connects with logic, reasons, and evidence. For example, in order to convince someone to stand up from their office chair once every hour, we might provide statistics to show a significant correlation between sedentary lifestyle and death resulting from cardiovascular issues or other chronic diseases. One can infer that sitting for too long is harmful to health. Ethos relates to the projection of the speaker including professional appearance or reputation, whether one is trustworthy or not. Considering the warning related to sedentary behavior, if the speaker looks like a domain expert, such as a medical doctor or a researcher, the statement would be more convincing. Pathos engages the audience's imagination and feelings. Common techniques that can trigger audience emotions include use of imagery, metaphor, and stories. These topics will be discussed in more detail in the next two chapters.

The idea of persuasion has been exploited by many theorists in modern times. With the advances of other media forms, the mode of persuasive communication is no longer limited to language. The tremendous changes in societal structures also liberate persuasion from public speech to everyday communication. Today, we might persuade a meat-eating friend to try a vegetarian dinner by sending them photos of varied colorful plant-based meat dishes as well as positive reviews from many non-vegetarian customers. The organic interactions between traditional media and social media even bring about collective persuasion. A long-running singing reality television show recruits dozens of inexperienced young perform- ers each season. They are to be shortlisted and trained by veteran coaches, before going head-to-head with each other as groups and as individuals. Finally, ten final- ists compete in the season finale.[2] In fact, the most valuable reward for the partici- pants is not the championship but opportunities of presenting a good show that also can be seen and appreciated by the social media audience. This kind of television show piggybacking on social media indirectly causes people to believe that pas- sion, dedication, and authenticity are more important than winning a game. Persua- sion of today can take place in disguise of varied fashions.

The study of persuasion has also been updated. Roland Barthes is one of the lit- erary theorists in the last century looking at the contemporary rhetoric on different media. In "Rhetoric of the Image" (1964), he analyzes how an advertising image generates multiple messages. According to his structuralist approach circa the early 1960s, he sees "rhetoric" in terms of signification and connotation, or in other words, different ways or levels of representing and conveying meaning. An image can convey meaning through multiple channels, namely, its caption or printed texts for the linguistic message, perceptual elements like colors or shapes for the visceral message, and culturally connotated elements like colors of the product for the sym- bolic message. Brian Sutton-Smith, in *The Ambiguity of Play* (2009), examines the rhetoric of play. He regards rhetoric in the modern sense as "a persuasive discourse or an implicit narrative" adopted to persuade others of the underlying belief (8).

Sutton-Smith uses this definition to discuss varied types of implicit narratives behind play in our culture. Rhetoric here has become even more indirect and subtle, yet surrounding us in everyday life. Other rhetorical scholars include Kenneth Burke (1969), who prominently addresses and declares the ubiquity of persuasion. As his famous quote puts it, "[w]herever there is persuasion there is rhetoric. And wherever there is 'meaning' there is 'persuasion'" (172).

The post-structuralist movement started in the mid-1960s dismisses that meaning is conveyed from the author to the reader. Instead, the reader can always interpret a message differently from the author's intent. The term rhetoric, in general sense, becomes concerned with how to effectively compose a work to shape the possible interpretations by the reader or the audience. Ian Bogost takes on this sense of rhetoric and argues that successful persuasion depends on effective expression, but does not necessarily lead to effective influence, at least not directly. Bogost is interested in the rhetoric of a new media form – computer games, and coins the term "procedural rhetoric" in *Persuasive Games* (2007) to refer to the practice of persuading through processes in general, and computational processes in particular (3). According to Bogost, persuasive games should embody arguments or discourses that simulate the ways real-world systems work. By playing a persuasive game, players virtually experience different processes and outcomes in the real world. They can follow the simulations and infer the rules governing the real world, which is the meaning of a persuasive game that players are supposed to make. For example, the once popular simulation game series SimCity circa 1990s allows a player to experience the process of urban planning. For steady population growth, one must allocate resources (such as land and taxation money) to balance the different needs of their citizens (such as utilities, healthcare, education, leisure facilities, businesses, and others). The simulation processes enable players to be aware of the many competing factors that make a city tick. Although one should not assume the game faithfully reflecting a real-world society, it makes us think about the relationships underpinning a real city.[3]

Bogost's thesis emphasizes persuasion through processes and thus criticizes the outcome-oriented approaches in using computers and related technologies to influence people's attitudes or behaviors, or what B.J. Foggs (2003) calls Captology (Computers as Persuasive Technologies). Fogg interprets "persuasion" in stark contrast to the contemporary sense of rhetoric. Rather than shaping interpretation, Fogg aims at changing behavior. He proposes an "elegant" behavior model, which can allude to a few well-known psychological theories (e.g., Theory of Planned Behavior). Fogg's model relates motivation to ability, whose magnitudes are presented, respectively, along two perpendicular axes of a graph. High motivation and high ability increase the likeliness to perform a behavior, yet triggers are still required to start the action.[4] Motivation is based on reasons for taking the action, which can be attributed to persuasive arguments. The more insightful part of Fogg's model is the suggestion of different types of triggers, which are useful for designing persuasive technologies to trigger behavioral change. Other scholars attempt to avoid the negatively connotated term "persuasive" and rename their approach as "behavior change support technologies," which aims to form, alter, or reinforce attitudes or behaviors "without using coercion or deception."[5] However, the term "persuasive technology" (as seen in many academic papers) is still loosely linked with behavior change or interventions.

In summary, the term "rhetoric" of today mainly refers to the arrangement of a work of expression or discourse, regardless of its modes of communication and media forms, that attempts to shape interpretations. Influence on beliefs, attitude, and behavior can be an indirect or extended consequence. Meanwhile, "persuasion" puts influence, particularly behavior change, on center stage. In short, rhetoric is about causing someone to believe, whereas persuasion focuses more on causing someone to change.

A Very Brief Introduction to Design

As persuasion can be found wherever there is meaning, it is not bound in the domains of literature, media, and communication. In the design field, people sometimes also speak about meaning. A designer might consider whether adding an option to a product is "meaningful" to the target customers or users. In the case of the automobile MINI revamped in 2001, the manufacturer provided a wide range of color options. The classic Union Jack pattern on the hardtop is especially meaningful to the original Mini owners or lovers, because such design passes down the classic Mini "spirit" to the new generation. The meaning here is associated with one's memory.

Sometimes, customers or users are also concerned with something beyond the individual, for the benefit of society or nature. They might find an environment-friendly product or service meaningful. For instance, the FREITAG bags, with the first prototype in 1993, are made with retired yet still durable truck tarps. The cutouts from each unique truck tarp render each bag distinct from others in color and pattern. Customers choose this product, not only for the distinct style but also due to the recycled material and the statement behind. The meaning emerges from the story of repurposing discarded tarps, giving each of them a second life, as what the company website says. In short, meaning is abstract, but it can be grounded in the stories connecting to a design.

Design may generally refer to any plan or visualization, or the art or action of producing it, that shows the arrangement of the features of an artifact before it is made. The artifact can be something in a three-dimensional physical form, such as a building, a vehicle, a piece of furniture, a utensil, a toy, and a board game, or in a two-dimensional media form like a poster, a film, and a video, or in varied digital formats, including a website, a mobile application (i.e., app), or a computer game. Practicing designers focus on the plan or visualization, but they cannot actualize their ideas without collaboration with other professionals. Designers need to discuss with structural engineers for a building, communicate with manufacturers to produce a household appliance, or work with software engineers to launch an app. Building on the knowledge and experience related to these tangible artifacts, people in the design field today, including professionals and academics, start to elevate design to the level of dealing with the less tangible, such as what they call "product-service systems" or online-offline customer/user experiences.

Those who want to advance the knowledge of design, mostly the academics or researchers, have been working hard to develop more theoretical constructs. They are concerned with frameworks and guidelines for varied aspects of design, including processes, methods, and approaches. Some well-recognized frameworks for design processes include Stanford's design thinking process (five steps: empathize, define, ideate, prototype, and test) and the British Design Council's Double Diamond model (four phases: discover, define, develop, and deliver). Although these frameworks present what looks like a linear process, actual application often requires iteration. For instance, the prototyping and testing steps in Stanford's model are inherently iterative. Testing might reveal pitfalls of a design idea, which inform changes in the next iteration of prototyping. Also, these frameworks are not mutually exclusive. Actual design practice always requires organic mixing of two or more of them. For example, divergent thinking as stressed by Double Diamond is useful in the ideation step of Stanford's model to generate more possibilities, and then the former's convergent thinking can guide the prototyping and testing steps in the latter to identify what should be selected. Other than frameworks that guide the design processes, there are many suggested methods for different purposes (e.g., data collection, idea generation, evaluation) at different stages of designing.[6]

Approaches refer to different ways of designing, which typically lead to different possible deliverables that reflect the values prioritized or the beliefs upheld. In general, designs are intended to meet target users' needs or wants. The design process commonly starts with user study employing methods like surveys or interviews, or occasionally co-design workshops involving target groups as participants, followed by iteration of prototyping and user evaluation. This is the widely practiced user-centered approach, which prioritizes users' satisfaction.[7] Other approaches can prioritize very different qualities. For example, Bill Gaver and colleagues advocate ludic design,[8] which aims to engage users in a kind of playful experience just like playing puzzle toys (e.g., Tangram), watching plot-twist movies, or photo-hunting wildlife in the safari. Ludic design stays open to interpretation and allows users to make multiple meanings (in analogy to arranging Tangram pieces differently for different figurative forms). By engaging with ludic artifacts, users yield certain, largely unexpected, benefits (e.g., Tangram helps players develop spatial composition and visualization skills).

Design Semantics and Rhetoric

While ludic design stays open to interpretation, alternative approaches attempt to limit users' interpretation for different purposes. In *Pretense Design* (2019),

Per Mollerup presents a category of designs that manipulate appearance to direct people's attention and shape their interpretation in varied ways, including over-statement (e.g., high heels), understatement (e.g., girdle), simulation (e.g., wigs), and hiding (e.g., camouflage). Mollerup, built on product semantics, draws analogy between pretense design and language. Product semantics, or design semantics, is an area of design knowledge that deals with how meaning, from functional to non-functional (e.g., symbolic or metaphorical meaning), emerges between designs and users.[9] In design semantics, a designer not only informs a user how to use a product but sometimes also suggests why one uses it. The primary reason is typically its usefulness and usability (i.e., easy to use).[10] Other reasons can be evocation of precious memories (e.g., MINI) or association with higher values like sustainability (e.g., FREITAG). These reasons constitute multiple levels of meaning. It is like an author composing a text that enables multiple interpretations (e.g., lyrics of the folk song "Where have all the flowers gone?").

As a design may allow multiple interpretations, it can also attempt to shape how people interpret it. This is design rhetoric. Instead of taking language as a medium of persuasion, design rhetoric regards artifacts as the medium with its message. Mollerup (2019) stresses that "pretense design is communication" (35). When someone uses a pretense design (e.g., wearing a hair piece to cover a bald spot), the design is intended to shape the audience's interpretation (e.g., making others to think that the wearer is younger than they actually are). That is, the user sends a persuasive message to convince others of something untrue. This kind of non-verbal persuasion, as the primary objective of pretense design, can also be a side effect of using a typical design. In fact, using a design often shows, either consciously or unconsciously, others how to use it. Wearing an outfit is an everyday example. When a person mixes and matches the top, bottom, and shoes and goes out, they present a visual image in the street that demonstrates how one uses those clothing items. It might appeal to others and cause someone to look for a similar apparel. Bringing your own tumbler for a coffee to-go or your own reusable shopping bag for grocery shopping is another ubiquitous example. A stylish tumbler or bag may remind others of the possibilities of looking cool while doing something good for the environment.

Advocates of design rhetoric believe that designers "give persuasive arguments a physical form" (36), which is to be experienced by users, as well as understood by others who see how the design is used. Further to that, designs of today are not limited to physical forms, but can also be digital. In contrast to the traditional modes of persuasion using language, I would call them tangible forms. By "tangible," I mean they can be, mediated or not, perceived by sight, hearing, or touch. Hence, tangible forms render something being easily seen or noticed. This is important for design rhetoric because the tangible design rhetoric must be visible to their intended audience (both users and onlookers), before they can be persuasive. With today's widely accessible digital media, the supposedly private use of a design can be easily made public. For example, home appliances are supposed to be used at home,

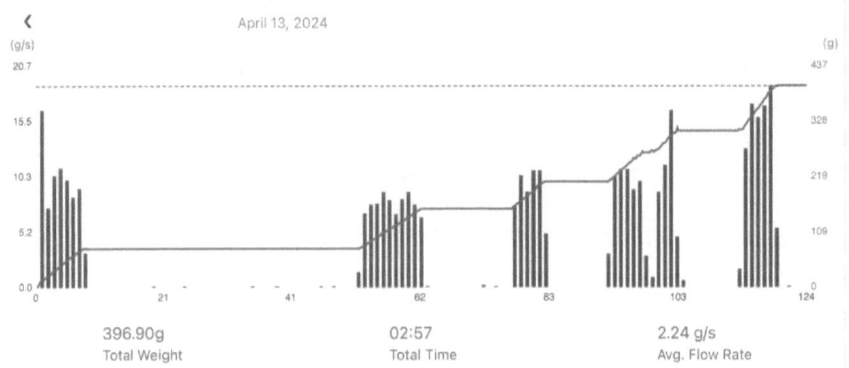

FIGURE 2.1 A screenshot of the mobile app connected to the smart scale, Acaia, presenting a weight-time graph with a curve showing the water flow and bars indicating the changing flow rate.

so that others should not gain access to people using them unless they are invited to someone's home. Yet, users might now share their trial uses of appliances on social media. Some smart appliances might come with mobile apps that allow sharing the use records on social media. For pour-over (aka hand-drip) coffee lovers, the smart scale, Acaia, connects to its app on the user's mobile phone and transfers in real time the continuous measurements. After a drip, the app presents a weight-time graph with a curve showing the water flow and bars indicating the changing flow rate (see Figure 2.1). To the user, the graph generated every day reveals the hand dexterity performance for steady and timely flow. One can put down a tasting note, check with the graph, and adjust the pouring skills and speed next time. A series of graphs accumulated across a period gives a sense of competence that one is able to tilt the kettle in nuanced ways as the graphs show varied slopes. The subtle variations in iteration are outcomes of one's intended choice of making changes. They satisfy two innate psychological needs, as informed by positive psychology, namely, the needs for competence and autonomy. On the other hand, the user can share the graphs on social media together with pictures of the cup and the tasting notes. One can choose to share with the pour-over community in hope of receiving comments or suggestions or share them with just one's own friends to let others know their dedication. A series of posts, seen by others, is like an evidence of the user's routine. This adds to the social proof, a term in social psychology, that this practice brings about possibilities.

The Psychology of Behavior

As mentioned, there are two sides of persuasion. One side is rhetoric, focusing on the arguments, expressions, or discourses that aim to cause someone to believe or to do something. From classical rhetoric to design rhetoric, persuasive arguments can be embodied in tangible forms. Digital media and technologies further facilitate presentation of a series of tangible outcomes resulted from the use of a design, which extends to the other side of persuasion—behavior change. Psychologists have been looking into what influences a person's behavior, including internal drives or external stimuli. According to a school of thought in positive psychology, people can make conscious choices between these competing factors. On the other hand, social psychologists think that there are two levels of cognitive processes, namely, automatic versus conscious,[11] and daily behaviors are often driven by the former.[12]

Positive psychology focuses on building humans' positive qualities, such as courage, curiosity, and self-regulation.[13] Self-Determination Theory (SDT), a leading literature in the field, systematically delineates motivation into different types along a continuum of how far a behavior is self-determined.[14] Intrinsic motivation, at the self-determined end, refers to doing something for one's own interest, enjoyment, or inherent satisfaction, like someone spending much energy and time on a hobby. Next comes extrinsic motivation, which includes subtypes where the main cause comes from the internal self or external systems. For instance, someone invests the time and money to learn driving and get a license primarily, not because they enjoy driving, but only for an anticipated road trip in Iceland. This is considered extrinsic motivation with a cause coming from the internal self. On the other hand, someone may learn driving at their spouse's request. This is considered extrinsic motivation with a cause coming from external systems. When a person is extrinsically motivated, one seems to be more concerned with the outcome rather than the process, resulting in less enjoyment before the outcome is actualized. Research informs that intrinsic motivation is associated with interest and excitement, leading to enhanced activity performance, persistence, creativity, and general well-being. To facilitate intrinsic motivation, SDT suggests three basic psychological needs to be satisfied, namely, competence, autonomy, and relatedness. When a person feels competent to perform an activity (e.g., through positive feedback), one becomes more self-motivated. The motivation is further strengthened if one sees choices and thus feels autonomous. Lastly, if there is secure support from others (e.g., care), the intrinsic motivation is even stronger. Consider again the smart scale Acaia. Using the device every day for pour-over can be self-motivated, leading to long-term engagement. When a user sees a weight-time graph generated after a drip, one can assess the performance, say on the steadiness and timeliness of water flow. If the review is positive, one can feel competent. Under most circumstances, however, there can be no obvious improvement after just a few trials. Additional meaningful rewards, in my opinion, can be suggested for a sense of competence. For instance, the curve in the graph can be transformed into a relevant image, such as a coffee plant grown in a pot, bearing flowers to represent the nuances one has made in a pour. The number of flowers indicates the number of pulse pours performed, and the size of a flower reflects the amount of water in the corresponding pulse pour. With today's generative AI technologies, this transformation is not

FIGURE 2.2   A composite image made from a photo taken by the author and an AI-generated image of a coffee plant.

rocket science (see Figure 2.2 for an example). A series of coffee plant images generated across a week or two constitute solid evidence of the work one has done to actualize possibilities. Moreover, each image is a result of one's intentional choice of pouring water over the coffee grounds in a particular fashion. This virtual but tangible reward represents one's competence and autonomy in pursuit of subtleties. If the user further shares the generated images on social media, resonance or appreciation from peers would add feelings of relatedness as well. According to SDT, the sense of competence, autonomy, and relatedness brought by the designed reward can enhance intrinsic motivation in users. This intrinsic reward (here echoing Wendy Wood's term as well, as the generated image is immediately and directly connected to the pour-over action) probably leads to better engagement and even more creativity (e.g., one might explore creative ways to generate different coffee plant images). In the next two chapters, I will discuss more along this trajectory.

Social psychology tends to consider forces from not only the inner self but also the environments. The former typically engages a person in conscious evaluation that involves one's beliefs about consequences (can be intrinsic or extrinsic motivators), important referents, and ability to control, resulting in intention to perform a behavior or not. The phenomena are summarized as Theory of Planned Behavior

(TPB) by Icek Ajzen (1985). For example, a person preparing for a journey would consider the anticipated scenery, food, or any safety issues (i.e., consequences), recommendations from friends or Internet celebrities (i.e., referents), and whether the required time and money are affordable (i.e., ability), before deciding to order flight tickets. Meanwhile, the environments are always at work in less noticeable ways. Features in an environment, as psychologists call "environmental cues," which can be objects or sensations that prompt a behavior without triggering a conscious thought. For instance, when a person sees a stack of paper towels next to the sink after washing hands, they tend to thoughtlessly grab them to dry their hands forgetting about the handkerchief in their pocket. Similarly, the vibe in a party may make one mindlessly pick up a drink or smoke a cigarette. John Bargh and Tanya Chartrand (1999) call this phenomenon an automatic environment-perception-behavior process. In contrast to planned behavior, automatic behavior is non-conscious. Sometimes, a behavior might be initially intentional, yet after frequent repetitions with consistent outcomes, it becomes automatic and even counter-intentional on some occasions. Commuting to work by public transit, for instance, can be on autopilot mode, which frees our minds for some high-level mental work, such as considering which colleagues should be involved for an upcoming meeting. The autopilot runs perfect until one day the metro changes the platform due to launch of a new line, and you might find yourself in a wrong train. This is the good and bad of habituation.[15] Other than conscious versus automatic processes, some psychologists come up with comparable terminology, like central versus peripheral, effortful versus effortless, slow versus fast, reflective versus impulsive, and others, which are collectively called dual-process theories.[16] Richard E. Petty and John T. Cacioppo (1986), in their Elaboration Likelihood Model (ELM), map a person's cognitive processes in response to persuasion onto two routes, namely, the central and the peripheral. When the person is motivated (intrinsically or extrinsically) and able to think (e.g., being distracted, exhausted, with prior knowledge or not) about the topic, thoughtful consideration (cf. TPB) takes place along the central route, which likely leads to a relatively enduring attitude change. The peripheral route is engaged when a person has little interest in the message or is less able to process the message, in the presence of peripheral cues (e.g., attractive, emotionally charged, novel; cf. environmental cues). The peripheral route likely produces a change too, but relatively temporary. Consider again the user of the smart scale Acaia who posts generated images on social media. These images can be regarded as tangible persuasive arguments to other people who see them. To a coffee lover who is not into pour-over (low motivation) or who just habitually goes to a coffee shop for convenience (low ability to think), the distinctive generated images still look attractive, which may connect him or her to the origin of coffee, that is, a tree. Thanks to their low motivation and low ability to think, this positive feeling toward using the scale for pour-over can only last for a short period of time. Hence, iteration of this action comes into play here. If the image poster iteratively posts to social media, each posted image has subtle differences and stays attractive to other social media users. The iteration at least persuades these followers in the peripheral route and hopefully engages the central route when their motivation and ability permit. In case a social media follower becomes interested in pour-over and thinks about trying it out, the cumulative serial posts of generated images can provide a solid evidence of the poster's persistent routine act, which likely strengthens the persuasive argument in the sense that someone has been doing it for a long time and doing great. Robert B. Cialdini, in his book *Influence: The Psychology of Persuasion* (2009), calls this phenomenon "social proof," which states that we "see an action as more appropriate when others are doing it" (88). The persuasive effect is most prominent for the social media followers who feel uncertain about the behavior, but then become convinced by others' evidence of performing it well out there. In this sense, the nuances and varieties in the serial, tangible outcomes resulted from someone's persistent, iterative actions as a routine, demonstrate the possibilities, which are not only appealing but also convincing.

Gamification or Narrativization?

This chapter explains why iteration of a routine act that yields serial, tangible outcomes can be persuasive. On one side of persuasion, rhetoric emphasizes the making of arguments, discourses, or expressions, in other words, the processes of persuasion. Design rhetoric offers foundation for embodying expressions in tangible designs that shape interpretations through manipulating appearance (e.g., skeuomorphism). Procedural rhetoric suggests utilizing computational processes in video games that engage re-interpretation of real-world systems through simulation (e.g., SimCity). A simulation or a game typically runs in an alternate reality, or what game theorists call "the magic circle," wherein rules and mechanics might be drawn from the real world,[17] but happenings in two worlds are separate (e.g., a new hospital built in a player's real-world vicinity yields no new building in SimCity). If a simulation or a game is designed to synchronize with part of reality, the simulated processes and outcomes are not only meant for one to experience and interpret but also intended to motivate. It extends to the other side of persuasion, that is, influence on behavior. For example, the mobile app Fortune City[18] encourages its users to record daily expenses and rewards with buildings in a virtual city. The categories of spending determine the types of buildings, such as recording a coffee bill contributing to a coffee shop, a cinema ticket to a cinema, and the like. This kind of design, strictly speaking, is not regarded as a game, because it no longer maintains the magic circle separate from the real world. Instead, people commonly call this approach "gamification," which is formally defined as "the use of game design elements in non-game context to motivate."[19] Game design elements include rules, goals, tasks, scores, levels, resources, constraints, and others. Gamification assumes that players' desire to win is a core motivator, which indirectly leads to success in non-game contexts such as learning, health, or workplace. Bogost, as a game and rhetorical scholar, prominently denounces "gamification" for its use of game elements only to change people's behavior. Such strategy downplays the rhetorical processes enabled by games and only sees games as a means to manipulate people for desired outcomes. More importantly, gamification has been employed by businesses and organizations in its most reductive form, using only basic elements like points or badges (symbolizing a player's level) to stimulate participation among target customers or even colleagues.[20] Some people call this simple and appealing approach "pointsification." Points or badges can be easily applied in any non-game context, say motivating learning, team management, or health interventions, because they are just context-free markers that measure progress. Yet, this context-free nature renders them extrinsic—they are externally imposed systems separate from the original activity domain. Recent research shows that extrinsic incentives have implications on motivation and performance. Elisa D. Mekler and colleagues (2017) compare giving points, badges, leaderboards, or no incentive to participants in an image annotation task. Results show that all three basic elements have no significant effect on intrinsic motivation and promote performance only in quantity but not quality. As extrinsic incentives, these basic elements easily direct one's attention away from the activity to only winning the game. Participants thus try to tag more images to earn more rewards and forget accuracy. The findings also support Edward L. Deci and colleagues' earlier study of motivation (1981). It empirically shows that direct competition (i.e., the fastest, the winner) has a significant negative effect on intrinsic motivation in solving interesting puzzles. In summary, gamification seems to receive criticisms both from rhetoricians for its reductive use of games for motivational purposes, as well as from psychological researchers for its inclination to extrinsic motivation.

To enhance intrinsic motivation, some persuasive technology researchers suggest using meaningful stories in line with users' goals and interests. Drawing on an ethnographic study of a massive multiplayer online role-playing game (MMORPG), World of Warcraft, Amon Rapp (2017) provides a review on gamification for behavior change and suggests design strategies for better utilizing game elements. The strategies include enveloping objectives of performing a behavior in a "narrative framework" (i.e., a story world). Similar implementations can be found in many gamified habit-building systems, such as Zombies, Run! that puts users in a story world with zombies. A user thus has to run in the real world and accomplish different tasks in the story world, like searching for healing potions. Meanwhile, Rapp's suggestion also reminds that there should be a variety of storylines satisfying different tastes of different people. Some people might be fans of science-fiction thrillers and like running away from zombies, but others might prefer puzzle-solving detective stories. It would be perfect if a system provides a user with options of different story worlds for framing one's goal and allows building one's own storyline by putting together the chosen tasks. This vision opens up a lot of design possibilities but also brings questions. A system might provide different story genres for users to choose from. Yet, how about the storylines? Among many known narrative structures, what plot types should be considered? Are there any common patterns among different storylines in different story worlds of different genres?

Scott Nicholson (2015), in response to the common practice of extrinsic gamification, proposes the notion of "meaningful gamification" that promises to help users find their own reasons for engaging with an activity (4), thus enhancing intrinsic motivation in them. The proposed strategies include creating stories for participants to integrate with real-world settings. He argues that a narrative, as a sequence of linked events, allows one to see the connections from past, through present, to future (7) and facilitates informed decision-making in real life if it mirrors part of reality. Meanwhile, stories, like games, can be a distraction from reality. Hence, the created story should be analogous to the real-world setting, providing noticeable hints for one to connect a story event with a status in reality. In this sense, the zombie story in Zombies, Run! gives its users a reason why one has to run, but it can be too immersive during use because of the run-along audio storytelling. Would stories like this be too distracting to users from the original goals in the real world? When and where should a user be reminded of the connections between those two worlds? How can a meaningful analogy be built to connect and balance the two worlds?

If the serial, tangible outcomes of iterating a routine act constitute a story analogous to the routine setting, the story development gives one a sense of competence (ability to develop in the story) and autonomy (making different choices). If the user shares a story update with others, they can feel relatedness through peer feedback, while others see the evidence and may become interested as well. In addition to intrinsic motivation, the story can also be designed in line with the user's vision in the routine and interest, thus making a metaphorical rhetoric through pathos. More importantly, there are correspondences between developments in the story and the routine. Hence, one can connect the current story event with the real-life status, making conscious choice for the next stage. In the next two chapters, I will further review and justify the value of using stories in terms of the imagery and imagination they brought, followed by discussion about the connection between a fictional story and reality.

3
IMAGINATION ACROSS REALITY AND FICTION

In expressive iteration, action results in something tangible that one can see or touch, which may elicit images on a person's mind. When browsing the photos taken on a trip, one can relive the moments as if revisiting those places. This attributes to our faculties of "capturing" immediate senses, "remembering" them as images in the mind, and "manipulating" these images in myriad ways. Sometimes, we might replay these images in a remembered chronological order. For example, browsing a sequence of photos may evoke one's mental images between the moments of taking those photos, which virtually walk the person through the trip again. Other times, we might combine mental images from different past experiences to form new images that are not authentic memories but speculative, hypothetical, or even counterfactual scenarios. For example, consider a person stayed in a hotel and took photos on its grounds. Unfortunately, the hotel has been destroyed by a recent fire. When viewing those hotel photos, they can imagine the harrowing scenes like hotel guests escaping their rooms in the fire, and wonder what it would be like being one of the guests there that night. These images are creative combination of one's direct experience of staying in the hotel and what one might have seen in disaster movies. Whether recalling a memory or imagining a scenario, putting together relevant images is like visualizing a story, which technically refers to a sequence of events. Consider iteration of an action. If the resulting tangible outcomes elicit a sequence of images that constitute a vivid story, one is easily absorbed into it. Image-laden narrative experiences prove to be effective in affecting people's beliefs and intentions.

If the serial, tangible outcomes resulted from iteration of a routine act constitute a story, the person performing the routine creates and tells their own story to express themselves. Photo-based food journaling can be an example. Someone trying to stick to a vegetarian diet starts taking photos of every meal they have,

DOI: 10.4324/9781003391449-4

be it served in the restaurant or self-cooked. A set of photos accumulated over two weeks and sorted according to the capturing time might show a pattern of increasingly more colorful dishes (veggie meals are typically colorful), which can be regarded as a successful story of becoming a vegetarian. Some might not totally agree with this assumption, because some commonly expected elements of a story are missing, say the protagonist, actions, conflicts, and conclusions. Others might argue that the protagonist is implicitly the person posting the food journal, and actions are reflected by the dishes, which are the outcomes of the person ordering food or self-cooking. The story is still ongoing and not yet finished. More importantly, this food journal expresses that it is possible to eat veggie meals every day without having the same dish twice! Then, one might want to ask the following questions. What constitutes a story? What are the essential components in a story?

In daily communication, we use the word "story" quite loosely. A teacher might ask a student being late for school, "Tell me your story today." In news and media, journalists are looking for stories. For marketing purposes, almost every brand now has the brand story presented on its website. It seems that a story is something to be told or presented. And the mode of "telling" is not limited to speaking or writing. We see how those in the creative industries use different artforms, like film, comics, or theater, to show and tell stories. Games are also another peculiar media form with indissoluble ties with stories. Many games are built on a backstory, ranging from board games like Cluedo to video games like Super Mario Bros. or Angry Birds. In role-playing or adventure video games such as Final Fantasy or Metal Gear Solid, stories even contribute to a larger part of the gameplay enjoyment. The interplay between games and stories has even sparked lively academic discussions between ludologists and narratologists over two decades ago.[1] Today, we have seen tremendous amounts of creative hybrid works enabled by ubiquitous digital technologies, some might think that drawing a clear line between games and stories seems no longer relevant. This chapter is not intended to re-engage that intellectual confrontation. Yet, for iteration to be expressive in some engaging ways like certain hybrid works do, it is relevant to rethink what stories and storytelling can be.

Storytelling—Imitation of Action

The classical academic references of Western narratology typically start with Aristotle's work. His *Poetics*, discussing the Ancient Greek tragedy tradition, introduces the most important element of a narrative—the plot, which is a sequence of events that make up the story in a structured way, including a clear beginning, middle, and end. Other elements that follow the plot are characters, who drive the actions in the story; thought of characters; language used by characters; rhythm of the story; and finally the spectacle.[2] His emphasis on the structure of narrative underpins later structural approaches emerging in the last century, including the French structuralist theory and the Russian formalist tradition. Seymour Chatman's *Story and Discourse* (1978) compares these approaches and comprehensively delineates the updated elements of narrative. Narrative includes a story, which is the content to be expressed; and discourse, which refers to the way of expressing the content. The content consists of events (actions and happenings) and existents (characters and settings). These components are interrelated. A character takes action in a setting (e.g., Mario in the video game jumps up to catch a power-up on the platform), and afterward the character may come across some happenings (e.g., the power-up on Mario expires and Mario loses the corresponding power). Chatman's model also covers the arrangement of the events and existents, as well as their manifestation in verbal, theatrical, pictorial, and cinematic modes. To apply the model to structuring possible stories conveyed by expressive iteration, actions and settings should be in the context of the focused routine, and the person practicing the routine can be the main character. As the person repeatedly performs the routine act, they generate a sequence of events. Happenings can be changes to the person due to the routine practice, and they may come slowly in reality. For instance, someone taking a 30-minute run every evening to improve physical fitness would need at least months to see noticeable changes in BMI (Body Mass Index). If the person could imagine being the main character in another story analogous to the routine, say, being Mario running to get power-ups, possible happenings can be seen more noticeably and regularly because the power-ups caught along the platform change Mario's appearance in an instant. The person's running routine can be structurally mapped with Mario's journey in terms of the narrative elements, including the main character (the person vs. Mario), the act of running, settings (e.g., a jogging trail vs. the platform in the video game), and happenings (slow, subtle vs. regular, obvious). Among the narrative elements, action is the nexus between the routine and the imagined story, because both sides share largely the same action. Other elements, to a certain extent, also demonstrate correspondences in the analogy, but their similarities are relatively loose (e.g., the person might not dress like Mario, the trail in reality is unlikely to be as straight as the platform in the video game). Indeed, not only does Aristotle consider event sequences the most important element of a narrative, but he also prioritizes actions in events. To him, storytelling (mainly of tragedy, a genre in the focus of his writings) is "imitation of action."[3] In ancient Greek theatre, a performer on stage might use voice and gesture to mimic a character and effectively convey an event in the story, even though the costumes and stage settings might not fully project the fictional scenario. The audience surely knows the stage is not the battle field in Odyssey, but one could fill in the background noise, soldiers, fire, smoke, and the like by imagination. The essential role of action in storytelling becomes most prominent in mimes. The audience, for example, sees a mime performer pull an invisible rope. They cannot see it, but they can "perceive" it by imagination. In fact, the trained performer also imagines there is a rope in the hands. Consider again the person practicing an evening running routine, without any particular technologies, they need to do what the mime performer does, imagining there is a power-up on the trail that can be picked up. To sum up, all stories are made up of events, which include actions and happenings. Storytelling is the imitation of event actions. To both the audience and the performer, successful imitation entails imagination.

From Imitation to Imagination

Since action is key to both a routine and the stories, it can be the vital link connecting a routine to an analogous story. Performing a routine act that resembles an action in a story is similar to "telling" the story to oneself or others, provided that other narrative elements such as characters or settings appear to facilitate the imagination. Things like costumes, props, and stage settings can support performers' actions in theatrical storytelling. Yet, it may not be easy to manifest these supporting elements in the contexts of daily routine. Fortunately, today's accessible mixed-reality technologies like augmented reality (AR) and location sensing enable the manifestation of these elements. Consider again the running routine. When the person runs along a trail, catching a power-up can be simulated by a smart watch that vibrates and earphones that ding. The smart watch might even display an animated graphic of running Mario and show the change in his appearance. The runner can also choose to run on a paved walkway that looks like the platform in the video game. The mix of the physical settings and the virtual items allows one to easily imagine oneself being or running with Mario to catch power-ups during an evening run. The imitation of the act of running, together with the mixed physical-virtual reality, as perceived by the runner, yield an imaginative story of running in tandem with Mario in the real world.

This imitation-perception-imagination process has been touched on by thinkers and theorists, as early as Aristotle. In another of his work De Anima (On the Soul), he differentiates three kinds of perception.[4] The first kind refers to what philosophers call sense data, like patches of different colors and shapes, sounds with a pitch, or tactile qualities such as vibration, which can be largely measured and simulated today. The second kind of perception occurs when the perceived looks like something. When we see a rudimentary hand-drawn animation of a matchstick figure, we can tell it looks like a cat elegantly walking. This perception already entails imagination. We compare, very spontaneously, the matchstick movements with how cats walk as we have seen in daily lives. The third kind of perception requires prior knowledge—we recognize a talented comedian mimicking an infamous politician by recognizing their well-known gestures and voice. Aristotle describes imagination as "that in virtue of which an image occurs in us."[5] Imagination produces images when there is no immediate perception, although the images can be largely based on perception in the past. Imitation also generates images, which are intended copies. They seem to be two kinds of images. Images resulted from imitation physically exist or have once existed, including mimes, pictures, or animations, while those from imagination only exist in one's mind. In other words, perception is a cognitive process that deals with images, arguably across the matter-mind boundary, from imitation to imagination.

Imagery and Imagination

The word "image" has an etymological root related to both "imitation" and "imagination." "Image" is in fact a slippery concept that requires a dedicated field of study in the humanities, what is called "iconology." W.J.T. Mitchell in his book *Iconology* (1986) presents a categorization of images horizontally. The left side includes physical images like pictures, statues, designs, and projections. As far as theatre is concerned, I think postures, facial expressions, and voices should also be included. The right side of the categorization sees mental images, which are what we see in dreams, memory, or fantasy. In the middle, there are perceptual images, which refer to sense data, that is, the first kind of perception as defined by Aristotle. More importantly, drawing on Ludwig Wittgenstein's critique, W.J.T. Mitchell tends to argue that the perceptual process from physical images to mental images is reversible in principle. That means the two kinds of images should be regarded as "images" in the same category (18).

Rudolf Arnheim in *Visual Thinking* (1969) describes how this reverse-perceptual process can be possible in art practice. Artists are used to producing physical images (e.g., a drawing of a cat) from mental images (e.g., memories of cats). While drawing, one would constantly review what is on paper and simultaneously adjust the lines or shapes until it looks right. The output is one single physical image, but it is unlikely that it is based on one single mental image. Instead, it is often a combination of many mental images of cats, such as what a cat owner sees every day, what one remembers from the television, or a cartoon character. It is natural to combine mental images, because they are typically incomplete and dependent on what the mind selectively remembers and recalls. Arnheim analogically describes it as "a complete body partially revealed by a flash light" (105). Hence, as an artist composes a drawing of a cat, they might catch mental flashes of their pet's short hair and long tail, how a cat sits or walks on a tree as they recall from a television documentary, and the dot eyes from a cartoon. Mental images, like physical images, can be detailed and specific in some parts (e.g., short hair, long tail) meanwhile blurred and generic in other parts (e.g., its true body form is unclear as it is covered by fur, but its body should be shorter than its tail). These characteristics inform structures of the mental images. Its body can be in any shape; its skin can be of any color. As long as it is depicted in short hair with a tail longer than its body, it can be recognized as a cat. Structures sharpen mental images on essential properties while blurring other details, like reducing human figures to skeletons in physical images, which can be useful for imaginative thinking. For instance, the mental image of a person running can be structured and presented in an animation of a matchstick figure on a flat surface. The animation can be precise, but the matchstick can be substituted by the skeleton and flesh of any character, which can be put in any place. This presents the structure of an action (running) in relation to an unidentified agent and setting, constituting an event in a story. This structure allows easy substitution of the matchstick figure with Mario as he runs on a platform with mushrooms.

The above structure of the mental image of running can be an example of what Mark Johnson (1987, 28–30, 171), George Lakoff (2012), and others call "image schema," a basic event structure in cognitive science that can be matched with compatible structures in another domain. In *The Body in the Mind* (1987), Johnson reviews major theories of imagination including Immanuel Kant's perspective, and identifies the schema as a procedure of imagination for producing images that belong to a concept (155). While a concept specifies the characteristics of an entity for it to fall under that concept, an image empirically provides an instance of experiencing those characteristics. A collection of related images co-visualize the characteristics of the concept. For example, our daily or recurrent experiences of dogs (e.g., four legs on the ground, big nose, round ears, wiggling tail, barking) result in many mental images, whose common basic features share an image schema, based on which we can imagine many examples of dogs. In the same vein, as we experience, remember, and even retell events in our daily lives as stories, we have many mental images of actions or happenings, which involve people, objects, and places. If these images are structurally organized and put in order according to some narrative schemata, we can use them to imagine new stories of particular kinds. Consider the running event again. We might have seen or experienced many comparable scenarios other than Mario's running in the video game, such as someone on the street running to catch a bus, an athlete jumping over hurdles on a running track as shown on television, or survivors in a horror movie running away from zombies. What are the common basic structures that can be seen in all these related images? Based on the narrative structure discussed so far, we could start with looking at the action-related aspects, followed by happenings, people, objects, and places. All scenarios see the act of running but vary in nuances. Some running toward a moving goal (bus), some running without a goal or definite direction but away from a moving threat (zombie), and some running from a clear starting point and toward a static goal while overcoming obstacles (hurdles) along the way. The running character may come across happenings such as being stopped by a static obstacle (hurdle) or a moving obstacle (zombie). The image schema might include a starting point, a goal (moving or static), a path (straight or winding), and obstacles (moving or static). Based on this schema, we might imagine a scenario consisting of a moving goal, a turning path, and moving obstacles, such as someone running for a bus, which turns at the junctions, and there are other pedestrians on the street.

To sum up here, a story is a sequence of events consisting of actions or happenings. Storytelling is imitation of the events, particularly actions. Imitation requires performers to imagine being in a hypothetical scenario. To spectators, perceiving imitation also entails imagination, because imitation can never be perfect even supported by costumes, props, and settings. From imitation to imagination, we deal with images of event components (actions, objects, places, etc.). These images can be organized according to their common basic structures or schemata. Action images can highlight the agent's body responsible for the acts and the movements, while leaving the objects open. Object images can show how agents can act on them or happenings may occur to them. Combining action and object images from different experiences makes it easy to follow the schemata. Putting together different event images is imagination.

Consider that a person performs a routine act (e.g., running routine) and simultaneously sees artificially produced images that follow, such as other characters (e.g., Mario), objects (e.g., a power-up), or surroundings (e.g., a treasure box). The person then plays the roles of both the performer and the spectator. While acting and perceiving in the real world (e.g., running past a mail box on a trail), one also sees images from the virtual world, which elicit alternate, imaginary events (e.g., Mario running past a treasure box and catching a power-up). Images from both the physical and the virtual can be selectively mixed, if they are structured by compatible schemata (e.g., running past a fixed object). The blend is a mixed-reality event (e.g., running with Mario past a treasure box on a running trail) wherein some components are added (e.g., Mario) or substituted (e.g., the treasure box replacing the mail box). A sequence of similar mixed-reality events constitute a story demonstrating expressive iteration. This kind of story is not fact or history, but fiction.

The Psychology of Imagination

Imagination, as a kind of cognitive process, has been broadly studied in psychology. M. Taylor (2011) describes imagination as a capacity to mentally depart from current time and place to think about the past (e.g., recalling the honey-moon trip), plan or anticipate the future (e.g., preparing for the new born baby), take another's perspective (e.g., being in an awful family situation faced by your close friend), create fictional worlds (e.g., making up excuse to decline a proposal), become absorbed in narratives (e.g., reading a novel), and consider what does not exist (e.g., living on Mars). The common activities of imagination include pretend play, narrative absorption, mental time travel, and counterfactual thinking.

Jerome L. Singer (1999) defines imagination in more technical terms and relates it to images. It is the ability to "reproduce images originally derived from the basic senses" as memories, fantasies, or future plans. These images include not only pictures and conversations but also remembered or anticipated smells, touches, tastes, and movements. These multisensory images can be reshaped and recombined into new images. For example, one might imagine revisiting a city but with a new family member this time, combining images of the familiar places, food, and transportation with that of the usual behaviors of that new member.

Lev Vygotsky (1991) is arguably a pioneer in studying imagination and creativity in childhood and adolescence, which he believes are more productive periods than adulthood regarding creative imagination. He differentiates creative imagination (like fantasy) from reproductive imagination (like memory). Creative imagination is "a visual reworking of concrete sensory images." It manipulates concrete images originally from immediate reality, abstracts them, and constructs a new concrete form of images. For example, one might have tried many pizzas before, and the concrete images might be abstracted into common attributes and basic ingredients including crust thickness, size, flour, cheese, tomato, rosemary, salami, and the like. One could mentally construct a concrete image of a novel pizza, together with images of fruit tarts, in an oval shape with plenty of green cherry tomatoes and blueberries. The intermediate abstraction of images is comparable to Arnheim's partially detailed images or Johnson and Lakoff's notion of image schemata. They enable production of new images.

In short, imagination is the "reworking" and "recombining" of concrete multi-sensory images, yielding new concrete images. The insights into the human ability to mentally reproduce images that no longer exist before their eyes have inspired many research questions. Like Wittgenstein's critique on the separation between physical and mental images, psychologists also ask whether mental images affect human thoughts and feelings to the same extent as physical images do. Under what conditions, would the effects of imagery be maximized? Craig A. Anderson (1983) asked 93 college students to imagine either themselves, a friend, or a disliked acquaintance as the main character performing a series of target behaviors (e.g., blood donation, taking a trip). To enhance the mental imagery, subjects were also asked to sketch these behavioral scenarios in cartoon-panel form. Before and after the cartoon task, their intentions concerning each of the target behaviors were assessed on Likert-point scales. Results show that (1) imagining oneself performing a target behavior produces corresponding changes in intentions, (2) such changes increase with frequency of imaging, and (3) such changes only occur when imagining oneself as the main character. Other researchers look into the relationship between imagining a hypothetical scenario and its perceived likelihood to occur. W. Larry Gregory and colleagues (1982) asked subjects to imagine themselves winning a contest (positive event) or being arrested for a crime (negative event). In each case, subjects reported that after imagining the events, they subjectively believed those events were more likely to occur. The effects on likelihood estimates also extended to behaviors. The researchers also asked homeowners to imagine themselves utilizing a cable television service, and subjects were subsequently more likely to subscribe to such a service when given the opportunity to do so weeks later. Steven J. Sherman led a study (1985) that provides more evidence on the relationship between imagination and perceived likelihood, which is mediated by ease of imagery. They asked subjects to imagine contracting a disease. For some subjects, the symptoms described were concrete and probably experienced. For others, the symptoms described were less familiar. Subjects who imagined contracting the disease with concrete symptoms estimated a higher likelihood of actually contracting it. The effect of imagination on likelihood estimates thus appears to be mediated by whether the imagery is familiar and easy to imagine.

Overall, imagining experiencing an event makes one believe more strongly that the event would occur, that is, increasing the perceived likelihood, which is mediated by the ease of imagining. Imagining being the main character performing a behavior and imagining frequently can effectively change one's behavioral intention and even actual behavior. Mental imagery affects individuals' beliefs about the likelihood of an event (e.g., heavy snowfall might snap tree branches) and one's intentions to take action and prepare for it (e.g., adding artificial support to stake a tree), not less than its physical counterparts.

Mental imagery can thus be used for varied purposes. For instance, it is applied in the context of exercise and sports. Athletes use varied kinds of imagery to enhance performance in competition.[6] An athlete can use imagery to mentally practice specific skills or rehearse strategies. This kind of imagery is cognitive-oriented and shown to be associated with improved performance of motor skills. One can also visualize one's specific goals (e.g., winning the tournament) as a reminder or imagine being excited or resilient in the game overall. This kind of imagery is motivational and found to increase self-efficacy, or one's beliefs in their own ability to effect the desired outcomes, leading to improved overall performance.[7] The successful use of imagery in enhancing sports performance provides ground for the use of imagery in constructing events other than sports competitions in pursuit of self-directed goals. The technique, sometimes called "mental simulation," is technically defined as the imitative representation (cf. imagery) of real or hypothetical events, including replays of past events, rehearsals of possible future scenarios, fantasies, or a combination of them.[8] Compared with the general use of mental imagery, mental simulation allows one to experience the actions and emotions in a series of events with a causal relationship between them. This is not unlike experiencing a story. There are two types of simulation, namely, process-focused or outcome-focused.[9] Process simulation emphasizes causal actions necessary for an outcome and the action-outcome links, while outcome simulation emphasizes the end results.[10]

Studies show that process simulation is effective in promoting behaviors that approach a desirable goal (i.e., approach goal), such as studying for a high grade,[11] physical activity for a strong body,[12] blood donation for a sense of well-being,[13] or daily fruit consumption for good health.[14] Highlighting the positive action-outcome links seems to be motivating because it prepares one for the necessary challenges or unwanted feelings associated with the action (e.g., getting used to the boredom of studying, or familiarizing oneself with where to get fruits and how they look) in pursuit of the benefits. Outcome simulation, on the other hand, can be useful for prevention of behaviors by just imagining potential positive outcome of stopping a behavior (still approach goal), like the pleasant moments after reducing alcohol consumption[15] or smoking cessation.[16] The process of stopping a behavior usually involves no action, and outcome simulation suffices. In some contexts, imagining undesirable outcomes of a behavior (i.e., avoidance goal) can be used to alter it. In smoking cessation, for example, many people successfully quit because they want to avert a current negative state like chronic cough (i.e., cure goal) rather than preventing a potential problem (i.e., prevent goal).[17]

In summary, imagination includes mental manipulation of concrete multisensory images into new images. The existing or once-existed images are abstracted and recombined into the new images. An individual can combine images of currently or previously experienced scenarios into new scenarios, whether they are past possible events, future likely events, or fantasy events. When a person has acquired and organized more images for the mental library, one becomes more able to produce images of a new event by combining existing ones, mentally experiences the new event more often, and believes more strongly that it is likely to occur. Hence, one can set a desired goal and imagine going through the processes to achieve outcomes for enhanced self-motivation and improved performance.

Some people may lack concrete sensory images in their mental library for constructing particular "unseen world," especially those involving multimodal senses.[18] It is so-called the "unimaginable" that one could not even draw on similar past experiences for a believable mental picture, for instance, an urbanite without much traveling experience could not imagine themselves living in the mountain. Today's advanced visualization technologies, for instance, so-called virtual reality (VR), AR, or extended reality (XR), can provide artificial sensory images that simulate hypothetical scenarios and facilitate imagination. They typically involve using varied sensors that detect changes in the environments (e.g., ambient lighting) or users' behaviors (e.g., body motion), connected microcontrollers that transmit tracked data in real time (e.g., via Wi-Fi), and microprocessors that render and present sensory stimuli (e.g., humanlike voice) to simulate a scenario (e.g., a virtual agent talking).

With VR technology, one can create fully immersive virtual environments that provide concrete sensory information of an imaginary world, including visual details, sounds, proximity, and even tactile feedback, as first-hand experiences. The one-time simulated experiences of hypothetical situations have been found effective in enhancing self-confidence,[19] reducing negative stereotyping,[20] motivating one to prepare for the future,[21] and changing one's attitudes toward the nature.[22] Although a participant experiences such realms just once, some effects may persist for a few weeks. Maybe it is because one has acquired "unseen" images from VR that enable them to mentally simulate the hypothetical situations again afterward. Other approaches include using AR that superimposes computer-generated images of a part of a virtual world or item on a person's view of the real world, commonly through a mobile phone with a built-in camera or an VR/AR headset, or any ambient displays or projections that blend images of a virtual setting into a physical environment. They can provide repeated experiences of the hypothetical scenario in one's daily life. The mental simulation can be even more frequent and timely. Hence the effects on behavioral intentions can be more prominent.[23]

The Psychology of Fiction

The idea of mental simulation focuses on imagination of past events, future events, or hypothetical scenarios that are close alternatives to the facts (e.g., taking a new challenge, putting oneself in another's shoes, imagining the future self). VR or AR can stimulate fantastical imaginations that are detached from reality, such as transforming oneself into an animal or experiencing supernatural power. These kinds of imaginations are not based on reality and are closer to fiction. It not only mentally immerses one in a hypothetical situation but also mentally transports one to a fictional world. Psychologists call it "mental transportation" or "narrative transportation," which is interestingly linked to literary theory.

When listening to a fairy tale (e.g., "The Frog Prince"), reading a novel (e.g., *The Lord of the Rings*), reading a comic book (e.g., *Dragon Ball*, *The Amazing Spider-Man*), or watching a science-fiction movie (e.g., *Planet of the Apes*), we of course know it is unreal, largely made up, or just fiction. Yet, we tend to suppose that the story really happens somewhere, probably in alternate reality or another domain or dimension. We try to understand and buy into the logic in this story world (e.g., people can learn and practice how to cast magic spells) to be fully absorbed into the story. We are willing to believe it, though temporarily, for the sake of storytelling. Regarding this literary phenomenon, the English poet Samuel Taylor Coleridge coined the famous term "suspension of disbelief." He asked his reader for "willing suspension of disbelief for the moment that constitutes poetic faith."[24] Hence, we do not challenge the cognitive capability of Sherlock Holmes or question the source of unlimited energy of the Hulk. Poetic faith is a kind of imaginative belief.

Richard Gerrig (1993) is arguably the first psychologist prominently studying the experience of reading a story. He likens the mental processes of reading to traveling. A reader, like a traveler, is transported some distance from his or her world of origin. In the journey, one becomes cognitively inaccessible to some aspects of the original world and hence puts them aside. One opens their mind to new ideas. When returning to the original world, one is somewhat changed by the journey. In other words, a reader becomes less aware of or temporarily forgets some real-world facts, mentally travels to another world, and takes away new information or knowledge from there back to the real world. Gerrig's work suggests that absorption into a story can affect a reader's beliefs. The effect of narrative transportation has been empirically studied by other psychologists. Melanie C. Green and Timothy C. Brock (2000) developed and validated a transportation scale to measure one's extent of absorption into a story, including mental imagery (e.g., how easily one can picture the events, how easily one can picture oneself in the scene of the events, how vivid the images of the characters are), cognitive attention to the story (e.g., how much one wants to learn the story's ending, how the ending could have turned out differently, how difficult it is to leave the story behind before finishing it), emotional involvement in a story, as well as the lack of awareness of one's surroundings when reading a story. In their experiments, subjects were asked to read a story about a murder committed by a psychiatric patient in a shopping mall, and then self-report their opinions or perceptions with respect to some elements related to the story, like the frequency and likelihood of attacks in public places, as well as measuring one's extent of mental transportation. Results show that higher level of mental transportation is correlated with beliefs that are consistent with the storyline. More importantly, whether the story is labeled a fact or fiction does not affect the extent of transportation. In summary, fictional narratives are persuasive. In addition, the effects have proved to last for weeks, indicating that beliefs changed by fictional narratives are integrated into daily life.[25] In other words, one has brought new knowledge from the fictional world back to the real world.

In psychological terms, Gerrig also draws on a dual-process model (that I have mentioned in Chapter 2). Corresponding to automatic and fast thinking, Gerrig's terminology is "unsystematic." For conscious and elaborate thinking, Gerrig's model supposes a kind of planning system that deals with reality. According to Green and Brook (2000), mental transportation onto a story world and absorption into a story take place more easily in the first kind. The reader turns off the action-planning systems and just accepts information without judging. When we sit in the cinema to watch a movie, we do not plan to act. So, we turn off the action-planning systems, and we do not critically check every twist and turn in a story against reality. However, if one plans to act, one must mentally simulate the act by testing it against reality to see if it really works. This reality-test inevitably disrupts narrative transportation. Norman N. Holland (2008) thus raises the issue of control in so-called "interactive storytelling" that allows the reader or player to continuingly choose one path among many options. Once the reader or player considers the options, he or she turns on the action-planning systems, which trigger reality-testing and reduce the extent of transportation. Studies using VR or AR, however, seem to provide evidence for the counterarguments. Participants in VR not only consider different options but also act out their own narratives in a virtual world. In a study by Jakki O. Bailey and colleagues (2015), for example, participants were put in an immersive virtual setting looking like a shower and were asked to wash their right arms in the virtual shower. A participant assigned to the experiment condition can see their avatar (a digital graphical representation of himself or herself) standing outside the shower window. One can access the virtual water tap to control the temperature and flow of the virtual water. For every 15 seconds of hot water flow, the avatar would eat a piece of coal with a crunching sound and cough with vibration feedback, which are supposed to exaggeratedly represent the energy consumption behind using hot water and prompt one's reflection. One can control the amount of hot water to be used and how much coal the avatar has to eat as a result. This is of course an unreal, fictional scenario yet loaded with vivid images and intense sensations. Before and after the experiment, each participant's use of hot water for hand washing in the physical world was tracked. Results show that participants seeing coal-eating images vividly used cooler water afterward to wash their hands in the real world. Interactivity and control granted to participants are unlikely to reduce their absorption into the virtual event and its persuasive effect.

I have also conducted a study using AR to create an illusionary experience of smoking in physical environments for influencing one's intention to quit smoking.[26] We designed and developed a smart ashtray in remote connection with a digital picture frame. Participants who were habitual smokers were asked to use the smart ashtray in one of their daily smoking sessions for five consecutive days. When one starts using the smart ashtray for smoking, a digital picture, which is personalized to the participant based on his or her favorites or priorities (e.g., soccer match, family photo) collected from a pre-trial survey, would get increasingly dusty. If the participant stops smoking and puts the ashtray back in place, the virtual dust would be slowly washed away by virtual rain. One can choose to put out the cigarette and start clarifying the picture of personal importance, or light another cigarette and watch the dust build up. Participants were asked to report their thoughts daily, on a scale of one to five, regarding three aspects: their intentions to quit smoking, vividness of the consequences of smoking, and perceived connection between smoking and the picture gathering dust. Results showed that the subjects reported increasing levels in all three aspects day by day, indicating one's increasing absorption into the virtual-physical hybrid bizarre event with corresponding changes in intentions, even though action-planning is involved.

The above studies support that when we are in control and plan to act in a fictional event, transportation may not be affected, and the persuasive effect is still prominent. We know the event is not real, but the experience of affecting the event through action, or technically the sense of agency in the story, enables us to mentally link the unreal to reality. We physically act on the tap differently to reduce virtual coal-eating. We stop the physical act of smoking to let the virtual dust gradually be washed away. Hence, experiencing agency in fictional narratives may activate more elaborate thinking than traditional storytelling, but it does not necessarily make a person challenge if the narratives are real. Instead, it establishes new connections between the fictional world and the real world. A person will extend their scope of thinking to cover the two connected worlds, because in both worlds they see something that they have concerns for.

4

CONNECTION WITH NEW INTERPRETATION

Iteration of a routine act, if deliberately designed and augmented, can regularly generate tangible outcomes that elicit images and form a story. By repeating the act in nuanced ways, a person may affect how the story unfolds. The ability to affect the outcomes in a story prompts action planning, which involves considering and visualizing different possible paths. A person thus makes connections between the physical routine acts and the possible storylines. However, this mental leap from the original routine to fictional outcomes can be distracting. Imagine someone who starts an early morning running routine to prepare for an upcoming marathon. If they imagine themselves running with Mario (the main character in the video game Super Mario Bros.) with the support of a smartwatch app that displays the character and delivers through earphones sound effects like those in the game, they might become so absorbed into the Mario story and so overly concerned with catching power-ups that they forget to maintain proper strides and breaths. In the beginning, they started the running routine to train muscle strength and cardiorespiratory endurance, so that they have what it takes to complete the marathon. This is the original reason for performing the routine. When new reasons are added for one to stick to a routine, those reasons should be aligned with the original ones.

To build stamina for long runs such as a marathon, one needs to maintain a steady pace and heart rate. To encourage such behavior, the duration of the temporary power boost after catching a power-up can be made contingent upon a runner's pace and heart rate. When these metrics are kept steady, the power boost will last longer. Another correspondence can also be drawn between these metrics and the coins that Mario collects. When the person's pace and heart rate become unsteady, Mario would carelessly drop a coin. Linking the story elements and happenings that Mario comes across with the desirable performance metrics of a routine is a cognitive process of connecting the story to the routine. With these mental

DOI: 10.4324/9781003391449-5

connections, the person not only pays attention to achieve the desirables but also sees how their performance intervenes the story development. If the story ending (e.g., Mario successfully reaching the castle and saving the princess) is also aligned with the original reason for performing the routine (e.g., one successfully crossing the finish line in the marathon), the person is absorbed into an actual-virtual hybrid story (e.g., running with Mario in the marathon and saving the princess at the finish line) wherein one stays focused on the most relevant aspects from both worlds. In short, expressive iteration stimulates the imagination of a fictional world in connection with the real world. It does not distract one from the routine but rather helps them focus on its extended cognitive network.

Connections in the Mind-Brain

The human mind-brain, through extending neural pathways, is able to handle a lot more connections than just those within one world. Today we know that the brain consists of billions of neurons, which are nerve cells that transmit information among each other with electrochemical impulses. Each neuron has an elongated portion known as axon (as a transmitter) that passes signals along to another neuron. Dendrites (as receptors), on the other hand, receive signals from other neurons. Between an axon and a dendrite exists a small gap called synapse. When an electrical signal from one neuron reaches the synapse, it triggers the release of chemical substances into the gap, which then binds to the receptor on the neighboring neuron, transmitting the signal and allowing for communication between groups of neurons, called nodes. A chain of interconnected nodes forms a neural pathway from one part of the brain to another. During the formation of neural pathways, axons of neurons grow, and connections between neurons are made at the synapses. Our daily experiences, including repeated practice or exposure to new stimuli, can, respectively, strengthen existing synapses to acquire proficiency or rewire neural pathways to adapt to changes (aka plasticity).[1] It follows that regularly experiencing ability to affect outcomes in a fictional or virtual world through physical acts would establish new neural pathways for learning and remembering new reasons for performing a routine (e.g., running with Mario to get more rewards). If the new neural pathways are in line with related existing pathways (e.g., running at steady pace and with steady breath), that is, the new and old pathways can run together in the neural network without any conflict and the need of rewiring in order to resolve the conflict (that implies unlearning or forgetting of existing knowledge about the real world), the hybrid physical-virtual experiences can deepen the existing real-world knowledge, while building new knowledge that extends to the fictional world. Based on this narrative on a neural level, we can articulate on a cognitive level that the new reasons emerged from expressive iteration should be aligned with (i.e., without conflict), or technically speaking, analogical to the original reasons for performing the routine, so that the person can stay aware of the real-world vision and simultaneously getting engaged with the fictional outcomes.

While it is hasty to equate neural links in the brain and concepts in the mind, studies enabled by recent advances in brain imaging techniques (e.g., functional magnetic resonance imaging) have found strong associations between neural activity in certain brain regions and analogical reasoning.[2] When a person more often performs analogical reasoning and other related cognitive operations, the corresponding brain regions are more engaged, and the related faculties can be more developed. The potential of the mind-brain is huge with practice.

Analogy between Routines and Stories

Analogy, generally speaking, is the comparison between two things that are similar in some way in order to explain or illustrate a point, idea, or concept. It helps people understand or learn something unfamiliar or intangible by drawing a connection between it and something more familiar or tangible. For example, a good high school teacher would explain electricity in analogy with water flow. While water flows from high places to low places, like from a cliff down a waterfall to a pool beneath, electricity flows from a point with high voltage to one with low voltage, like from the positive terminal of a battery to its negative terminal through an external circuit. The bigger the difference in voltage, the stronger the electric current. Similarly, the bigger the altitude difference between two points, the faster the flow of water. The similarities between electricity and water can be observed in their common behaviors, rather than material properties like being waves or atoms. The analogy connects electricity and water by highlighting their commonalities. Students understand voltage and current in electricity through the common patterns that can be found in the flow of water. Besides, it is exactly their differences that make the analogy work. Water flow is visible in daily life and familiar to most students, while electricity is not often visible except under extreme situations such as lightning. I suggest that the power of analogy would work between a routine and an imaginative story. The processes and outcomes of the routine acts may not be always arousing. Yet, the story events can be more engrossing. Analogically connecting the two highlights both the common as well as different aspects of both worlds for a person to see. The crux lies in whether and how a connection can be drawn.

The above electricity-water comparison shows that analogy emphasizes similarity not literally between the objects. Electricity is not like water in appearance or material, but they share common behaviors in moving from somewhere of high potential to low potential. This kind of commonality is technically known as relational or structural similarity, like FLOW (material, high, low) in cognitive science, an interdisciplinary field of study integrating psychology, neuroscience, linguistics, philosophy, and others about human cognition. A well-received theory of analogy in cognitive science is Dedre Gentner and Arthur B. Markman's structure-mapping theory (Gentner 1983; Gentner & Markman 1997). It claims that analogy involves mapping of relations between objects or their attributes (e.g., electricity in a battery and water in a reservoir), rather than mapping of objects or their attributes (e.g., battery is cylindrical vs. reservoir is irregular), from a base domain of knowledge (e.g., what we know about water flow) onto a target domain (e.g., electricity). It also specifies the so-called structural alignment process for drawing correspondences in the mapping, requiring that matching relations must have matching arguments (e.g., number of objects or object attributes and their types in a relation), and all matches are one-to-one correspondences. In short, analogical reasoning requires alignment of relational structure. For the above electricity-water analogy, one needs to look for common relations of objects, namely, electricity and water, or their attributes. A suggestion can include HIGHER THAN (x, y) and FLOW (a, b, c) wherein HIGHER THAN () is a relation taking two arguments and FLOW () is a relation taking three arguments. In the water domain, the relations will be:

HIGHER THAN (high-altitude, low-altitude)
FLOW (water, high-altitude, low-altitude)

In the electricity domain, the relations will be:

HIGHER THAN (high-voltage, low-voltage)
FLOW (electricity, high-voltage, low-voltage)

The two domains now have common relations and matching arguments, which are structurally aligned, although the objects or attributes in the matching arguments are different. They are called "alignable differences," referring to differences connected to the common structure. In contrast, "non-alignable differences" are those independent of the matching structure, which, in this case, include the different paths of water flow and the electric circuit. When people compare two things by analogy, they tend to focus on alignable differences rather than non-alignable differences. Markman and Gentner (1993) have conducted experiments and collected evidence to support this claim. In these experiments, subjects were asked to list the differences between similar word or picture pairs (e.g., police car and ambulance). They readily pointed out mainly alignable differences (e.g., a police car carries suspects to a police station, while an ambulance carries victims to a hospital). Meanwhile, participants who were asked to list the differences between dissimilar word or picture pairs (e.g., kitten and magazine) tended to point out non-alignable differences (e.g., kittens have flesh and bones, while magazine is only paper). Moreover, subjects tended to list twice as many alignable differences between similar pairs than non-alignable differences between dissimilar pairs. Results of these empirical studies seem to support a hypothesis—a comparison between two things that are similar in some ways actively triggers structural alignment leading to a focus on matching relations, which also foregrounds alignable differences. In fact, this finding should make perfect sense to us. Most things in the world are dissimilar and not comparable. We only concern ourselves with those similar pairs, and their subtle differences really matter to us.

It follows that if an imaginative story is analogous to a routine, a person can perform structural alignment and draw correspondences between them, leading to a focus on the matching relations as well as alignable differences, which connect the fictional outcomes and the original reasons behind the routine. Consider the running routine and the Mario story. The story structure that consists of actions and happenings in events discussed in Chapter 3 can be useful to our discussion here. In aligning the routine and the story, one can start with matching action relations, which typically take two arguments, namely, an agent responsible for the action and an object to be acted upon, like RUN ON (person, place), PASS (person, object), or BUMP INTO (person, object). Meanwhile, happening relations might take two or more arguments, including the character or object to be involved and the attribute to be affected, like UNSTEADY (someone, attribute) or LOSE (someone, attribute). In the domain of the running routine, the relations can be instantiated as follows:

RUN ON (person, street)— the person runs on the street
PASS (person, mailbox)— the person passes a mailbox
UNSTEADY (person, heart-rate)— the person's heart rate becomes unsteady
UNSTEADY (person, pace)— the person's pace becomes unsteady

In the Mario domain, the relations can be:

RUN ON (Mario, platform)— Mario runs on the platform
BUMP INTO (Mario, block)— Mario bumps into a treasure box called "question block"
LOSE (Mario, power)— Mario loses his power as it expires
LOSE (Mario, coin)—Mario drops and loses a coin

The relations in the two domains can be arranged and aligned in a table with the left column listing running routine relations and their arguments (one every row) and the right column listing Mario's relations and their arguments, as seen in Table 4.1. This is a result of structural alignment applied to the routine and the Mario story. A common relation shared between them is obviously RUN ON (), and the alignable differences connected to this relation include "person" versus "Mario," and "street" versus "platform," that is, the matching arguments. Other action relations that are not commonly shared by the two domains but can be virtually aligned by technology are PASS () and BUMP (). The former relates the agent "person" who passes the object "mailbox," while the latter relates Mario to the "question block" that he bumps into for a power-up. These two actions can co-occur, because they can "virtually" take place at the same time along the course of running with the support of latest location-based sensing technologies. The alignable differences here are also "person" versus "Mario," as well as "mailbox" versus "block." Similarly, some happening relations can be synchronized and virtually aligned.

TABLE 4.1   Structural alignment between the running routine and Mario's running

Running routine		Alignment		Mario's running
RUN ON (	–	common relation	–	RUN ON (
person,	–	matching argument	–	Mario,
street)	–	matching argument	–	platform)
PASS (	–	virtually aligned relation	–	BUMP INTO (
person,	–	matching argument	–	Mario,
mailbox)	–	matching argument	–	block)
UNSTEADY (	–	virtually aligned relation	–	LOSE (
person,	–	matching argument	–	Mario,
heart-rate)	–	matching argument	–	power)
UNSTEADY (	–	virtually aligned relation	–	LOSE (
person,	–	matching argument	–	Mario,
pace)	–	matching argument	–	coin)

Once Mario has caught a power-up, the power boost lasts until the user's heart rate becomes unsteady as measured by vital sign sensing technologies. The happening relation in the routine domain UNSTEADY () relates the user "person" to the changing heart rate, which triggers Mario's happening of losing his power as in the relation LOSE (). The two happening relations have again alignable differences of "person" versus Mario, as well as "heart rate" versus "power." Lastly, another pair of relations UNSTEADY () and LOSE () refer to the happenings where Mario drops a coin when sensors (such as a built-in accelerometer) detect unsteady pace for a period of time.

Expressive iteration helps people find new reasons for performing a routine by suggesting an analogous story. People become interested in the fictional outcomes, which should be properly designed and aligned with the original reason behind the routine. Analogy provides a systematic way of aligning the two worlds in terms of story structure. In applying structural alignment to story components including actions and happenings, the most curious part is how to identify the correspondence. Between the routine and the story, some actions are commonly shared, such as the act of running in the above running-Mario mapping. It is relatively easy to put together arguments for the common relation like RUN ON () that relates an agent to an object, although the arguments need to be properly put in place for the respective roles to match. Other matches can be less obvious. For instance, in the pair of action relations PASS () and BUMP INTO (), the person's act of passing and Mario's act of bumping into the block are compatible, because the two events can co-occur along the course of running as long as the person and Mario virtually running together. Yet, finding a match between a fixature in the street to be passed and the block to be bumped into for a power-up still requires some imagination. One might need to recall images of the street and visualize different possible objects that exist in it, such as a mailbox, a lamppost, and traffic lights. To align with bumping a block, passing a mailbox might be a better match than passing a lamppost or traffic lights, because the shape of a mailbox looks like holding something in it. This comparison suggests the correspondence, "mailbox" versus "block," in the relations PASS () and BUMP INTO (), respectively.

A Third Story: New Interpretation

In the structural alignment process, we imagine a person and Mario virtually running together in a real-world street. Possibly, the runner passes a mailbox, while Mario bumps into it for a power-up. We inevitably take a step further from analogically comparing two scenarios to selectively combining them into a third scenario. For each pair of matching arguments, either one argument or both are selected and projected onto the combined scenario. In the first pair of matching arguments of RUN ON (), both "person" and "Mario" are selected. In the second pair, only "street" is selected. The selective projection means that both the person and Mario run on the street. In the matching relations PASS () and BUMP INTO (), the first matching arguments "person" and "Mario" are again selected, while only "mailbox" in the second is selected. The combined scenario is that the person passes the mailbox, while Mario bumps into it for a power-up. In the first pair of UNSTEADY () and LOSE (), all arguments are selected, projecting a scenario where the person's heart rate becomes unsteady, and Mario meanwhile loses the power boost. In sum, structural alignment of the running routine and the Mario story leads to a selective projection of scenarios from two sides onto a third combined physical-virtual scenario wherein the person runs together with Mario on a real-world street. It is like we mentally transport Mario and his actions from the video game to the real world. That said, there are other possible selections resulting in different combined scenarios. For example, selecting "platform" instead of "street" in the matching relation RUN ON () would result in the transportation of the person into the video game; selecting and projecting "person" and "mailbox" onto the matching relation BUMP INTO () would outrageously have the person bumping into a mailbox. Some combinations can be even more ridiculous or irrelevant, such as projecting "Mario" and "heart rate" onto the matching relation LOSE (). It looks like structural alignment is a divergence process that opens up lots of combinational possibilities, and selective projection brings focus onto relevant and useful scenarios. The question is how to identify a meaningful scenario.

Analogy is typically used for the purpose of understanding, as in the water-electricity mapping, by projecting knowledge from a base domain onto a target domain. After structural alignment, projection always runs from the base to the target without much flexibility. When analogy is applied to connect a routine with an analogous story, the purpose is not to make it easy to understand but to generate novel interpretations. Structural alignment prepares story components, like actions, agents, objects, or happenings all in place from both domains. Projection thus goes from the two domains onto a third new story world wherein the person can see themselves affecting the fictional outcomes (e.g., maintaining Mario's power) as an integral part of pursuing the real-world goals (e.g., maintaining a steady heart rate). They are both reasons for doing the routine, becoming a new interpretation of it. This kind of selective projection from two inputs onto a third output is the core of another type of cognitive operation, conceptual blending, introduced by Gilles Fauconnier and Mark Turner (2002). Blending is the integration of two or more concepts into a new one, which often takes place on our minds in daily lives for varied purposes, such as reasoning, creative expression, and imaginative play. An illustrative example of blending raised by Fauconnier and Turner is "the boat race" between a multi-hull vessel *Great American II* sailing in 1993 and a single-hull vessel *Northern Light* sailing in 1853 on the same course from San Francisco to Boston, wherein the former tries to break the record of 76 days and 6 hours set by the latter. The race is not real, only exists in the mind of a sailing magazine journalist. To start this imaginative blend, we can structurally align the two events in terms of matching relations SAIL (*Great American II*, SF-Boston, 1993) and SAIL (*Northern Light*, SF-Boston, 1853), as well as POSITION (*Great American II*, nth day, position i) and POSITION (*Northern Light*, nth day, position j). Then, we can selectively project from both sides onto a third new concept, also a story, consisting of both *Great American II* and *Northern Light* sailing together, which allows us to see their relative positions and on which day the former taking over the latter. To make sense of putting both vessels together, this new story comes with new relations RACE (*Great American II*, *Northern Light*, SF-Boston) and BECOME (*Northern Light*, ghost), which are brought in from familiar background knowledge of "boat racing" and "ghost" in our culture. The import of knowledge frames, which consist of new relations outside the inputs, into the blend is called "pattern completion" by Fauconnier and Turner. Pattern completion adds new relations to the blend, re-organizing agents, objects, or settings, turning the new story into a boat race, and framing it as distinct from the inputs. In short, blending extends analogical mapping from structural alignment to include selective projection and pattern completion, generating new stories as novel interpretations.

In the blending processes, identifying additional knowledge frames and appropriate relations from those frames to re-organize the original components from the inputted stories are crucial to the generation of new stories. Knowledge frames, including the relations, serve to provide possible occasions wherein characters, actions, or objects, respectively, from two sides can co-exist in the third story. For the boat race blend, the knowledge of boat racing and the relation RACE (), typically involving at least two boats, are obviously good choice for imagining the two separate vessels being simultaneously on the same course. For the blend between the running routine and Mario's story, putting together the person and Mario in a "race" on the street might not be a good idea, because the person's goal of practicing running is not breaking a record, but rather maintaining steady pace and heart rate. The domain-specific knowledge of "running buddies" might be better here, which means getting a companion to practice running together. The knowledge frame [running buddies] is more specific to the domain of trail running and less common in urban running, that is, outside the input concept of the running routine. Importing the relevant relation BUDDIES (person, Mario) can tie them together in the new story wherein the person and Mario become running buddies when practicing running on the street. Whenever the person passes a mailbox, Mario purposely bumps into it for a power-up. That can be represented by a new spatial relation SIDE BY SIDE (PASS (), BUMP INTO ()) covering two original relations, and the two actions can co-occur. If the person's running pace turns unsteady, so does Mario's, causing Mario to accidentally drop coins. This connection corresponds to a new logical relation CAUSE (UNSTEADY (), LOSE ()). Other than this choice of pattern completion, another knowledge frame that can be considered for import is Mario's brother, Luigi, who appears in the video game in two-player mode. When they are running on the same platform, they have to be in close distance (i.e., within the width of the television screen). The relation BROTHERS (Mario, person) can be imported for the person and Mario to run together. Because they are brothers, Mario might sense the person's heart rate, which affects Mario's power duration. This can also be represented by the new relation CAUSE (UNSTEADY (), LOSE ()). In short, the import of new relations from external knowledge frames binds together the originally unrelated actions or happenings, turning each pair into matching relations, such as PASS () vs. BUMP () as well as UNSTEADY () vs. LOSE (). Table 4.2 summarizes the selective projection from the two domains, the running routine in the left column and Mario's running story in the right column. The pattern completion is also shown in the middle column, with the external knowledge frame and relations imported.

TABLE 4.2 Blending between the running routine and Mario's running

Running routine	Selective projection and pattern completion	Mario's running
	[running buddies]· BUDDIES (person, Mario)	
RUN ON (person, street)	→ RUN ON (person, street) RUN ON (Mario, street)	← RUN ON (Mario, platform)
PASS (person, mailbox)	SIDE BY SIDE (→ PASS (person, mailbox), BUMP INTO (Mario, mailbox))	← BUMP INTO (Mario, block)
UNSTEADY (person, heart-rate)	CAUSE (→ UNSTEADY (person, heart-rate), LOSE (Mario, power))	← LOSE (Mario, power)
UNSTEADY (person, pace)	CAUSE (→ UNSTEADY (person, pace), LOSE (Mario, coin))	← LOSE (Mario, coin)

With expressive iteration, a person explores new reasons for performing a routine (e.g., running practice) by first comparing it to an analogous story (e.g., Mario's story) and then blending them into a third imaginative story (e.g., running with Mario on the street). This third story consists of fictional events (e.g., acquiring power-ups) from the analogous story as well as the person's routine acts and intended outcomes (e.g., running with steady pace). For the person to see both the fictional events and the intended real-life outcomes as one set of goals, the third story comes with new relations that bind them together. These new relations are imported from knowledge frames (e.g., [running buddies]) typically outside of the routine and its analogous story. The third story is thus a blend of the two sides meanwhile being distinct from them because of its emergent structure and relations. Not only does the emergent structure include real-world goals and fictional events as new interpretations of the routine, but it also enables elaboration on the third story development. The latter is particularly important for expressive iteration. A routine entails iteration of acts, while an analogous story is a sequence of events including actions and happenings. The third story is then a sequence of blends between each routine cycle and its corresponding story event. For the running routine, the person runs on the street for 40 minutes every morning. They want to maintain their pace and heart rate steady in every run, which would enhance their stamina for the upcoming marathon. Yet, the progress would not be obvious without iterating the run daily for a few weeks. To add variety to the iteration, one might try a slightly different route every day, seeing and passing different fixtures, trees, or shops on the street, such as a lamppost at a junction, a tree with a birdhouse on it in the park, or a grocery store that opens very early in the morning. The variations are comparable to Mario's journeys on different levels with different themes in the video game. On one level, Mario runs on a grassland and might acquire a "mushroom" power-up that makes him bigger. On another level, Mario runs inside a cave and might acquire a "light box" power-up that puts a light in front of him, or a "feather" power-up that allows him to glide for a distance after jumping. When blending the running routine with the Mario story, each running route needs to be mapped carefully in accordance with Mario's journey. With the knowledge frame [running buddies] imported into the third story, Mario also passes the same street settings as the person does. Connections between the street settings and the types of power-ups can be drawn. When the person passes a lamppost, Mario might acquire a light box power-up. When he or she passes a grocery store, Mario might get a mushroom power-up. When passing a tree with a birdhouse on it, Mario might get a feather power-up. The third story features the person and Mario running side by side, and they come across the same items along the course. Connections between real-world items and fictional power-ups like those aforementioned can be thoughtfully developed for different cycles of iteration.

As the progress of practicing a routine is typically not obvious for a couple of weeks, a person sometimes needs intrinsic, immediate rewards or tangible progress indicators to maintain the momentum. To keep one interested and motivated, a third story from the blend of the routine and an analogous story should be further elaborated to reflect one's perseverance in the iteration. Regarding the running routine, the level-up game mechanics in the Mario's video game can be a straightforward match for the blend. For instance, in the first week, the runner sees Mario coming across power-ups only in the grassland (i.e., level one). In the second week, Mario starts to acquire power-ups in a cave (i.e., level two). This match is straight-forward, because the knowledge frame [level] is largely within the knowledge domain of "video games." Other than that, more creative matches can be found in the evolving styled representations of Mario in the video game. In the first week, Mario can be visualized in only low-resolution three-color two-dimensional pixel graphics. In the second week, Mario can have higher resolution and more colors that make him look like a cartoon with outlines and shading. In the third week, Mario turns into a three-dimensional computer-generated character. This is in fact a "meta" knowledge frame brought in the blend series, that is, the [game graphics] in a video game series. The newly imported relation from this frame can be ADVANCE (Mario, rendering quality), which can also be applied to the person's unseen progress as ADVANCE (person, stamina). On the one hand, a sequence of events are "happening" to Mario, that is, increasingly advanced technologies used to render him more finely, in each generation of the video game series. On the other hand, the runner's stamina is improved each week, albeit not readily visible or consciously known. Their weekly invisible progress is then paired with one visible change in Mario's rendering. As the knowledge frame informs that each change in Mario's rendering represents an improvement in visualization technologies, the runner also sees their stamina improving together with the increasingly advanced graphics. In other words, the visible changes in Mario can be perceived as the intrinsic outcomes of performing the running routine, because the runner and Mario are running together every morning. These visible changes become new reasons for the runner to keep practicing.

It is important to identify a sequence of consistent events in the third story and relevant relations that can bind together the cycles of a routine and the sequential events. The relations can be found within the existing knowledge frame like [level] in the domain "video games." More creatively, the relations can be imported from a loosely connected frame like [game graphics]. The crux is that the sequence of events, whether from [level] or [game graphics], features a pattern of forward movements along a direction, which is comparable to the runner's progress made toward their goals. Many stories describe a protagonist literally moving along a course toward a final destination for the purposes of saving someone, acquiring a precious item, or accomplishing a mission. Mario's story can be regarded as an archetypal quest. His goal is to save the princess. He moves on the "platform" of each kingdom toward the castle at the end, finishing one level and proceeding to the next. This forward movement also means getting closer to the final goal. A quest story like this can be an obvious option to be mapped with one's progress. The evolution of Mario's rendering in the video game features another metaphorical forward movement along a direction of increasingly advanced computer graphics. Whether the forward movements are literal or metaphorical, the pattern is still a common story structure.

Analogous Stories for Different Routines

In Chapter 3, we start with the most general definition of story as "a sequence of linked events." The links, to the least extent, can be inferred by the audience, if not stated explicitly. E.M. Forster, the English novelist, in *Aspects of the Novel* (1993), defines a story as "a narrative of events arranged in their time sequence" (21) and a plot with the additional "emphasis falling on causality" (60). He also offers a succinct and well-known example. "The king died and then the queen died" is a story, while "the king died and then the queen died of grief" is a plot. The latter explicitly points out the causal link between the two events. Yet, the former also invites its audience to infer some sort of relation between the deaths, which is likely to be due to deep sorrow or related emotional problems.[3] In other words, the readers or audience inevitably infer a causal link between the consecutive events in order to make sense of the story. And this inference "demands intelligence and memory," as Forster puts it (61). He describes the reader's interpretive processes. One sees events as either isolated or related to previous events. Some isolated events could be tentative, for the suspense in the plot. One would constantly rearrange and reconsider to look for new causal links. The processes thus require one to remember in order to understand. Brian Boyd in *On the Origin of Stories* (2009) takes a cognitive and evolutionary approach to analyzing the sense-making processes of stories. When a person grows up from infancy to adolescence, one advances the interpretation of stories from perception to inference, from description to explanation, from temporality to causality to long-term goals (150). When a story unfolds to reveal a new event, one would first retrieve preceding events from working memory. If one could not see any obvious link, one would try to recall other events already told in the story and stored in long-term memory. If a connection is still not found, one would turn to semantic memory that stores general knowledge (152). From the cognitive science perspective, general knowledge is stored as frames in the mind-brain, which are data structures like templates or formats for filling in missing information.[4] Hence, reading the extremely condensed story that the king died and then the queen died, one would turn to semantic memory to retrieve applicable knowledge frames like [marriage], [love], and [bereavement] to explain the event sequence. As mentioned in the discussion of analogy, knowledge frames contain relations that bind items or attributes. Oftentimes, the relations that link events are causal, like LOSS (queen, king), CAUSE (LOSS (queen, king), GRIEF (queen)), and CAUSE (GRIEF (queen), DEATH (queen)). Sometimes, causal relations could not be found. Given another context for the two deaths, the king and the queen were from different countries, and they did not know each other in person. The story telling the sequential deaths might just send the message that the two royal families coincidently lost their important members in succession. The events are just put together under the same theme. The knowledge frame one might invoke can be [coincidence] together with the relation SUCCESSION (LOSS (country A, king), LOSS (country B, queen)). L. David Ritchie in *Metaphorical Stories in Discourse* (2017) defines a story as "any recounting of a series of causally or thematically related events" (23). This definition provides a concise and inclusive perspective on the possible links in event sequences.

As a story is a sequence of (causally or thematically) linked events, it takes shape. Forster calls it "pattern" and provides examples of novels that take the shape of an hour-glass due to symmetry as seen in the plots. Gustav Freytag, a German novelist and playwright in the nineteenth century, proposed in *Technique of the Drama* (1900) arguably the most well-known model that prescribes the shape of a story, later called "Freytag's pyramid," which is also seemingly symmetrical. The pyramid figuratively represents the structure of a story, including five stages found in the plots of nearly all plays in his era, namely, exposition, rising action, climax, falling action, and conclusion. The exposition, represented as the first horizontal line preceding the pyramid, is the description of background information, for example, the backstory of Mario and why the princess needs to be saved. The rising action, represented as one side of the pyramid, refers to a series of increasingly significant events leading to the climax, such as the enemies sent successively by the villain to challenge Mario, and Mario's acquiring of different power-ups. The climax is at the top of the pyramid, which is the heightened moment of a story. It can be the direct confrontation between Mario and the villain. The falling action is on the other side of the pyramid, which is like the reverse of the rising action. For instance, the hero tries to undo what the villain has done. Here Mario has defeated the villain but is still trying to locate where the princess is locked up. The final stage is the conclusion, which shows the final outcomes of the story, including deeds that cannot be undone. Freytag's pyramid represents a symmetrical story structure. Yet, it is seldom purely symmetrical. The rising action is typically the longest and thus the major stage. The linked events in this stage produce "a progressive intensity of interest" to both the story characters and the audience, and they show "an enlargement in form and treatment" with "variation and shading" (128). If the events in addition show a repetitive yet slightly varied pattern, the rising action stage can be an ideal match with a person's progress in performing a routine. Thus, identifying an analogous story with iterative events in the rising action and mapping the successive events with one's progress in the routine provides a sense of episodic achievements in terms of the story outcomes. Such kind of rising action is particularly useful in the case of routines for goal attainment, like a running routine in preparation for an upcoming marathon, regularly practicing breathing exercises for mental wellness, or spending some time on reading every day for long-term intellectual development. This kind of routine is usually initiated by oneself, that is, with intrinsic motivation. Yet, the lack of visible progress or immediate outcomes within a period of time, say in a couple of weeks after the routine begins, can be occasionally demotivating. Blending the practice with an analogous story with iterative events in the rising action matches one's intention to move toward a goal.

Yet, not every routine is practiced for goal attainment. Many routines are performed, not because we want to achieve a goal, but rather to maintain our daily life. In other words, they are maintenance routines. For instance, we need to brush our teeth every morning, get dressed properly and commute to work, finish a quick lunch and go back to work, do laundry and other chores at home. Some people also have to regularly self-monitor vital signs like blood glucose level for chronic disease management. We typically do not expect ourselves to make significant progress in these routines. A person seldom sets a goal like doing laundry in order to be an expert in operating the washing machine, or doing blood sugar tests to be dexterous in the finger-pricking process. Instead of visualizing progress, one would be more pleased to see variety in these originally mundane routines. Expressive iteration could help here, yet an analogous story should consist of repeating events that result in a variety of outcomes under a theme. In Freytag's words, the events show "variation and shading" but not necessarily "an enlargement in form and treatment." The events "rise" not in size but in variety. A maintenance routine should be mapped with a story that features variety among its iterative events, just like Monet repeatedly painting haystacks in tandem with nuanced lighting at different times of day and in different seasons, the Jew in the "Good Samaritan" parable told in Bible repeatedly asking different people for help, or a player of Wordle repeatedly solving a different five-letter word puzzle each day and resulting in different color patterns. Events in a story of this kind are put together under a common theme, and thus thematically related. In the story, emphasis is put more on variation than progression. In Part II, we will start discussing different types of stories for different kinds of routines.

PART II

Case Study

In Part I, we arrive at the conclusion that iterative expression typically shows the following characteristics:

- Iterating routine acts result in a series of tangible outcomes, which function as records of progress or variations made by a person, satisfying innate psychological needs including competence and autonomy (Chapter 2).
- The serial, tangible outcomes are visible to other people, demonstrating possibilities as social proof that appeals to them (Chapter 2).
- The serial, tangible outcomes elicit images that enrich one's mental library and enable mental simulation of possible scenarios, which affect one's intention. (Chapter 3).
- The mental images form a fictional story. Transportation into a fictional world enhances one's beliefs related to the story (Chapter 3).
- Fictional stories may distract one from the original vision in real life. Fictional stories should be analogous to the target routines. Each fictional event maps with one cycle of a routine through structural alignment of repeating actions or happenings, as well as agents or objects (Chapter 4).
- Selective projection of actions, happenings, agents, and objects, respectively, from a routine domain or a fictional story, together with the import of external knowledge frames that reorganize the projected components, yield an imaginative third story as new interpretation of the routine (Chapter 4).

To construct expressive iteration around a routine, we need to identify a story with recurring events that are analogous to the routine, align the fictional events with the routine acts, and blend them into a third story. Appropriate alignments and blends between the story and the routine keep a person's focus on relevant aspects

DOI: 10.4324/9781003391449-6

from both sides. Meanwhile, different types of routines are motivated by different reasons and thus require different kinds of stories for mapping. Hence, Part II aims to characterize stories of different forms that can make routines expressive. In the next three chapters, we will look at the cases of iteration on three levels.

For self-initiated, goal-directed routines such as regularly doing exercise for fitness, progress made toward the goals (e.g., reducing blood cholesterol level or achieving an optimal body mass index) is typically non-obvious until a substantial period of time (e.g., three months) has lapsed. One's motivation can be better sustained by visualizing progress within a story where repetitive actions or happenings help the protagonist move forward along the event sequence. Stories with this structural trait can be found in many folktales across cultures, called "cumulative tales" or "chain tales." On the other hand, in practicing what I call maintenance routines such as flossing the teeth, doing laundry, or having quick and convenient meals during workdays, people seldom set a goal. Visualizing progress may not be major here. Sometimes, a person might want to add color or variation to each repetition of the routine. That can yield a sense of novelty, which is arguably another basic psychological need. Toward this end, one can map the routine with a story that involves recurrent events resulting in a series of varied outcomes under a common theme. It is like a series of creative work, such as Monet's painting series *Haystacks* and Japanese woodblock print series such as Katsushika Hokusai's *36 Views of Mount Fuji*. In Chapter 5, we review and analyze these works through the lens of analogous stories for mapping with routines wherein people prioritize either progress or variety.

For someone more concerned with progress, stories like chain tales are more effective in keeping them motivated. For those who prioritize variety, theme-based narratives should be the stories of choice. Routines performed for different reasons can be mapped with different types of stories accordingly. Yet, routines are typically long-running and may last for more cycles than the repetitions of events in most stories. Thus, in Chapter 6, I suggest the next level of iteration. Iteration of an analogous story with variations into sequels, which then form a series. As informed by well-known film series such as *The Terminator* (1984–2019) or *Indiana Jones* (1981–2023), chain tales can be iterated into episodic series. Meanwhile, many well-received woodblock print series in the Edo period of Japan were in fact followed by sequels. They also demonstrate another kind of series, that is, anthology. Episodic or anthology series of analogous stories can be mapped with long-running routines. To fuel people's momentum across successive stories in a series while practicing routines, two approaches are introduced in Chapter 6, with reference to the narrative devices, namely, the frame story and continuing storylines in serials.

Sequels, series, the frame story, and serials, can be made to strike a balance between introducing variety to the routine and implying progress, and thus, may encourage someone to carry on an expressive iteration. However, people sometimes still have to skip, suspend, or even quit a routine due to certain reasons. If a person, after a hiatus, resumes a similar routine, one might not remember the original analogous story or might no longer be interested in it and want a reboot. Others, oftentimes, just feel bored with the genre, theme, or motif of the story. Here, another level of iteration comes into play. Variants are versions of the same thing that differ in some respects but all share the common essence that can be called "nexus." Variants are common in folktales. Chapter 7 is a review on variants of a few exemplary chain tales suitable for mapping with routines. Strategies for generating new possible variants will be suggested, followed by demonstrations that update or personalize some tales based on individuals' preferences or needs. Variants of theme-based narratives, including Japanese woodblock print series and the traditional Chinese winter calendar charts, exhibit even more variations in styles of presentation or forms of representation. Some might see them as parallels or adaptations. Insights also can be drawn from these distant variants, which inform the formulation of new variants that are made possible by the latest mixed-reality technologies.

5

ITERATION OF EVENTS

People have different routines to perform in their daily lives. For different routines, there are different reasons. Sometimes, a routine is driven by a specific goal. A person may want to start a morning or evening running routine in order to prepare for an upcoming marathon, have oat or muesli for breakfast every morning to lower their blood cholesterol level, work out three times a week to build muscles, spend one hour every other day on a language course to prepare for their study abroad, or check their blood glucose level to monitor a chronic health problem. When practicing this kind of routine, one would like to know how much progress they have made toward the goal, for it satisfies the basic psychological need for competence as an intrinsic motivator. The progress seems to suggest how far from accomplishing a mission one is. Aligning and blending a goal-directed routine with an analogous story can produce tangible progress indicators as the fictional events get progressively intense.

Other times, very often indeed, people have to maintain a routine that is supposed to last for an extended period, without the intention of reaching a specific goal. Instead, these routines are just deemed necessary in our daily lives. Many of us have to commute to work and back at least several days a week. During workdays, we oftentimes need to have quick and convenient lunch meals. After work, we may need to do the laundry or other chores like cleaning and vacuuming. Most individual would regularly take a shower, brush their teeth, or better still, floss them. These routines might sound like what Frederick Herzberg called "hygiene factors" in his two-factor theory of motivation in the workplace, which are required to avoid negative effects like dissatisfaction but cannot enhance positive effects like motivation.[1] Herzberg uses hygiene as a metaphor for the workplace, eloquently describing the characteristics of this category of human needs. Yet, what I meant by maintenance routine sometimes can exceptionally bring about satisfaction or

DOI: 10.4324/9781003391449-7

pleasurable experiences. For example, a person enjoys walking their dog every day, but they might just get distracted or overwhelmed by other tasks like work and skip the dog walking routine every now and then. Some people are willing to build a habit of recycling materials that would otherwise be thrown away as trash. They believe in the benefits recycling brings to society or nature at large. Someone might prefer to walk the stairs instead of taking the elevator, not only for physical health but also to reduce electricity consumption and hence, carbon footprint. Thus, a maintenance routine is not motivated by a specific goal, but people still do it for some underlying belief or long-term vision regarding a lifestyle. In other words, maintenance routines are performed to maintain a lifestyle. With a lifestyle belief or vision on their mind, a person chooses to maintain the routines that are faithful to such belief, or necessary to support such lifestyle, although at times the routines could be mundane and unattractive (e.g., doing the laundry, recycling used plastics). Instead of making progress in maintenance routines, adding colors to them sounds more promising in satisfying the psychological need for novelty. Aligning and blending a maintenance routine with an analogous story can add variations to the repetition through varied fictional events.

In Part I of this book, I introduce the basic concepts of stories and storytelling. The definition of story starts with a sequence of events. Events include actions taken by a character (i.e., agent), happenings that a character comes across, or even an object. Storytelling is the representation of events, or more specific, an imitation of events, including actions and happenings. Narrative can be defined as structured storytelling, that is, organizing a story in a structure that includes components such as settings, characters, events, or relations. Plot refers to a narrative with an emphasis on the causal links between the events. One well-known story structure is Freytag's pyramid, which divides a common narrative structure into five stages. The main part is typically the rising action between the exposition and the climax, referring to the chain of related events that produce "a progressive intensity of interest" and show "an enlargement in form and treatment" with "variation and shading." This definition of the rising action matches well the priority of visible progress in practicing a goal-directed routine. The progressively intense and enlarged events in a story sequence can serve as tangible progress indicators along the course of routine practice. Meanwhile, the mentioned "variation and shading" in the rising action leaves room for variety, which is favored in maintenance routines.

Freytag introduced this model to describe the plays and novels circa the nineteenth century, which may seem dated. In fact, the rising action leading to the climax is still indispensable to many mainstream entertainment dramas of today, from the *Sherlock* television series (2010–2017), and the *Harry Potter* novel (1997–2007) and film (2001–2011) series, to many recent superhero movies. To the main characters (and the audiences indeed), an event sequence builds interest and heightens their emotional engagement (i.e., the rising action), as the story leads them to a direct confrontation with the big boss of the villains, such as Moriarty,

Voldemort, or Thanos (i.e., the climax). In fact, entertainment dramas still feature rising actions prominently. Moreover, the main characters usually have a goal. Sherlock wants to solve all the puzzles in a crime investigation. Harry Potter and his friends face threats and need to defeat the dark force. Superheroes are destined to save the world. The prominent rising actions and characters with clear goals suggest that this kind of entertainment drama can be a promising story group for mapping with goal-directed routines. However, routines are same actions performed over and over again. A story for mapping with a routine should also involve repetition of a similar action. Although entertainment dramas usually involve recurrent events, for example, Moriarty repeatedly challenges Sherlock to solve a case within a tight time frame, or Harry Potter repeatedly attends classes taught by a professor newly on board, the repeating actions oftentimes intertwine with other seemingly minor events to create twists for surprise and entertainment, like Sherlock noticing very subtle clues in earlier encounters to connect the dots, or Potter catching a new professor in suspicious act that actually misleads him and the audience. The highly engaging interwoven plots in these entertainment dramas would easily distract a person from the original goal of performing a routine if mapped with a daily routine. Among the commonly called "plot-driven" narratives, a promising analogous story for mapping with a goal-directed routine is preferred to have a linear and simple storyline. To sum up, the criteria for choosing an analogous story include: (1) the rising action is composed of repetitive events that show progression; (2) the main character has a goal; and (3) causal links between successive events are obvious. Interestingly, many folktales feature these characteristics. In particular, a subtype of folktales, called "cumulative tales," also known as "chain tales," provide very good references and patterns for imagining and crafting analogous stories for mapping with goal-directed routines. In this chapter, I will introduce a few exemplars and analyze their connections with different routines.

On the other hand, maintenance routines are not motivated by any specific goals. People practice maintenance routines to support a lifestyle. Oftentimes, they might find the processes repetitive and boring. One would like to see variations in the processes rather than progress. Mapping with maintenance routines requires analogous stories that are different from those criteria mentioned above. The rising action, characters' goals, and causal links may not always apply. As a story is a sequence of related events, the relations between sequential events are not necessarily based on causality. Sometimes, events are put together because they share a common theme (e.g., because they take place at adjacent locations in sequence, or because they involve similar objects appearing at different locations). For a story, the sequential events can be causally linked or thematically related to each other. In the latter, causality and the plot are not obvious. There may be causal relations among events, but the narrative structure is not built on them. The connections between events seem to be chronological, procedural, spatial, or even categorical. For example, Monet's *Haystacks* series can be regarded as a theme-based narrative that presents a sequence of different haystacks in the region at different times of day across seasons from the summer of a year till the following spring. Each painting, with a unique number, is an image or an imitation of an event, that is, a haystack having appeared in a field at a certain moment. The events, imitated by numbered paintings in the series, are spatially (in the fields of the same region) and categorically (using the same motif) related. Some might argue that this is not a story. Yet, from the general definition of a story as prescribed in earlier chapters, a sequence of thematically related events is considered a story. More importantly, these kinds of theme-based narratives are the perfect candidates for mapping with maintenance routines. We will discuss a few exemplary candidates of this kind in this chapter and connect them with potential maintenance routines.

Plot-Driven Narratives for Goal-Directed Routines

For aligning and mapping with goal-directed routines, such as running every evening in order to prepare for next year's marathon, having oat or muesli for breakfast every morning to lower blood cholesterol level, working out three times a week for building muscles, or taking language classes to prepare for studying abroad, an analogous story should fulfill three criteria: (1) the rising action prominently shows repetition and progression; (2) the main character have a goal; and (3) the causal links between repetitive events are obvious. A special yet not uncommon type of folktales, that is, chain tales, provide very useful references and exemplars.

Collecting Folktales

Folktales belong to folklore, closely related to and even mixed with, fairytales and fables. They are oral stories originating in a culture passed down from generation to generation. Fairytales may target primarily younger audiences. Fables usually have animals as the main characters and deliver moral lessons like parables.[2] They can be found in almost every culture. Although they originated in the oral tradition, some people collected and turned them into written texts. The earliest collection is arguably *Aesop's Fables* by a slave in ancient Greece,[3] including well-known short stories like "The Fox and the Grapes" and "The Ant and the Grasshopper." In the nineteenth century, scholars or researchers in Europe started to collect and publish folktales or fairytales known in their respective areas. The Grimm Brothers published *Children's and Household Tales* (1812–1863, tr. 1884), which is a German collection, including "Snow White" (ATU 709) and "The Fisherman and His Wife" (ATU 555). Russian poet and novelist Aleksandr Pushkin recorded folktales from his nanny, a peasant woman, and turned some of them into literary forms like verse from 1822 to 1828.[4] Pushkinian tales also include "The Dead Princess" and "The Fisherman and the Fish." The former is strikingly similar to the Grimm Brothers' "Snow White," and the latter is believed to come from the Grimm's fisherman tale.[5] Pushkin's publications of folktales, together with a few small collections by others, were followed by the famous and huge collection of primarily Russian and Ukrainian folktales edited by Aleksandr Afanas′ev (1st ed. 1855–1864, 2nd ed. 1873, tr. 1916). After Grimm Brothers' and Afanas′ev's success, other parts of Europe also saw publications of their own folktale collections. Joseph Jacobs edited and published the book *English Fairy Tales* (1895), which includes the well-known tale "The Three Little Pigs." Andrew Lang published a series of 12 *Coloured Fairy Books* from sources across expansive territories, even including stories from beyond Europe. For instance, the *Crimson Fairy Book* (1903) contains a supposedly Japanese folktale "The Stonecutter," whose plot is comparable to Grimm Brothers' version of "The Fisherman and the Fish." They belong to the same tale type in the Aarne–Thompson–Uther (ATU) classification system.

Classifying and Defining Folktales

With the flourishing numbers of folktale collections in that period, people have started to systematically categorize folktales. Finnish scholar Antti Aarne took tales from several major European collections and established tale types, each of which referred to a group of tales sharing a common character, such as a recurring plot or motif. He found that tremendous number of folktales across many different areas were surprisingly categorized into only a limited number (less than 1,000) of types.[6] For example, the type ATU 555, titled "The Fisherman and his Wife," refers to not only the tale collected by Grimm Brothers but also other variants like the Russian variant and the Japanese tale "The Stonecutter." In addition, American scholar Stith Thompson added motif descriptions to Aarne's index. Motifs for ATU 555 include "Wishes granted without limit," "Fish returned to water: grateful," and "tabu: making unreasonable requests."[7] While the ATU index provides a comprehensive organization of folktales across the world, criticism includes the difficulty of tracing a tale back to its type or categorizing new materials in appropriate types.[8] Moreover, plots and motifs sometimes lead to different types. For example, "Cinderella" involves "stepmother" and "stepdaughter" as motifs that fall into the class "magical task," yet its plot belongs to the class "magical helper."

Regarding recurring plots in folktales, the most widely known and highly respected scientific analysis is probably *Morphology of the Folktale* (1928, tr. 1968) by Soviet formalist Vladimir Propp. In *Morphology*, Propp takes a structuralist approach to studying 100 folktales from Afanas′ev's collection, separating the tales into component parts and comparing them according to their parts (i.e., "morphology," describing the relationship between parts and the whole) (19). He proposes a fixed series of 31 components, or "functions," each with some "varieties," that constitute all fairy tales in the samples. Propp's model can be seen as an extended, formalized system of studying folktales, an equivalent to Freytag's pyramid for dramas. In interactive narratology enabled by recent computing technologies, Propp's model has formed the basis of story generation. As Propp points out in another work, *The Russian Folktale* (posthumously 1984, tr. 2012), many definitions of folktales, like "implausible miracles," "invented narrations," or "made-up" events, are from the perspective of content rather than form (70–2). *Morphology* seems to be Propp's pursuit of defining folktales in a structuralist way.

Defining and classifying folktales has been a challenging task. Based on the Afanas′ev's collection, Propp (2012) summarizes a few categories, namely, tales about animals (including about objects, plants, or the basic elements), wonder tales (i.e., fantastic folktales), novelistic (or everyday) tales, and others (36–7). This categorization is of course not perfect either. For example, a wonder tale with fantastic elements typically involves talking animals as characters, which can be regarded as animal tales. Finally, Propp singles out a special category that is the most relevant to our discussion about mapping with goal-directed routines, the cumulative or chain-form tale, such as "The Turnip," "The Gingerbread Man," or "The Rooster Choked."

Cumulative (aka Chain) Tales

Thompson, when revising Aarne's index, was also aware that every kind of folktale has a very particular composition and style. He set aside ATU 2000–2399 to this category, formula or cumulative tales, for example, 2044 Pulling up the Turnip. However, a large number of typical cumulative folktales are scattered among other types, such as "The Stonecutter." Propp (2012) describes the particular compositional pattern featured by cumulative tales as "constantly increasing repetition of one and the same action, until the created chain breaks or unravels in the opposite, diminishing direction" (276). The statement reconfirms that a cumulative tale is a chain of events with the same action repeated, constituting the rising action, leading to the climax, and then the falling action. The emphasis on the repetition of action meets the basic criterion for mapping with routines. Each repetition of the same action in a cumulative tale usually adds some quantitative or qualitative items, leading to "gradual growth" or "piling up" along the chain. For example, in "The Three Little Pigs," the pigs' houses are built with progressively stronger materials, from grass straws, to wood, and then bricks; in "The Turnip" collected by Afanas′ev, every repetition of the pulling action involves one more person, with only the grandfather pulling at first, joined by the grandmother, the granddaughter, and so on. The "gradual growth" in structural strength or "piling up" of manpower demonstrates the most obvious feature of the rising action, that is, "enlargement in form or treatment." In many cumulative tales, the main characters have clear goals, and the actions are repeated due to direct causes. The peasant family in "The Turnip" wants to pull out the gigantic turnip, and they summon increasingly more members but repeatedly failing. The three little pigs want to fend off the malicious wolf, and they build a stronger house whenever the old one falls apart. The characters' goals are clear, and the causal links between repeated actions are direct. Many cumulative tales fulfill the three basic criteria of mapping with goal-directed routines, including the repetitive rising action, a clear goal, and direct causal links.

Meanwhile, characters in some cumulative tales have less well-defined goals. In "The Fisherman and His Wife," or its variants like "The Stonecutter," the main characters are repeatedly given a chance to make a greedier wish, gradually getting hold of more power. Although the repetitive rising action shows progression, and the causal links between successive wishes are obviously related to humans' insatiable desires, the main characters do not have clear goals in the beginning. In "The Flying Ship," a folktale from Eastern Europe, Ukraine, and Russia, as well as in Lang's *The Yellow Fairy Book*, the successive, recurring events seemingly happen without connection. The protagonist, typically portrayed as a simpleminded, looks for ways to build a flying ship without any plan. Fortunately, along the way, he repeatedly meets new companions with varied extraordinary powers, all useful in helping him tackle the impossible challenges ahead of him. The protagonist only starts with a wishful thinking, while repeatedly inviting super-powered strangers to board his ship, resulting in gradual growth of the magical team. Teambuilding is not part of the initial goal, and the links between repeated actions are loose. In fact, the order of the recurring encounters is not crucial and can be arbitrary. The crux lies in that the whole team was on board before the final stage. The above examples show that some cumulative tales, although made up of recurring events that form a prominent rising action, might not perfectly fulfill all the criteria of mapping with goal-directed routines.

Categorization of Chain Tales for Routines

According to the chain creation trait, Propp (2012) identifies two kinds of cumulative or chain tales (277). The first kind, as in "The Turnip," the gradual growth along the chain is due to main characters' clear motivation. The second kind, like "The Flying Ship," the reasons behind the serendipitous piling up are not very clear, probably not revealed until the climax unravels. For mapping with different routines, I suggest that chain tales can be further categorized. Although chain tales are primarily plot-driven, the main characters' goals and intended actions render a slight inclination toward the character-driven end. The categorization of chain tales dissects the inclination, in terms of whether recurring events are motivated by main characters' clear goals, and whether the piling up of event outcomes is intended by the characters.

The first category includes those tales in which event motivation is clearly tied with characters' goals, and event sequences are intended too, such as "The Turnip," and "The Three Little Pigs." The main characters' goals, being not yet achieved, stay as reasons and direct causes for repeating the actions with gradual growth in intensity that are supposed to help the characters get closer to their goals. This category is the most character-driven. These tales are ideal candidates for mapping with routines directed toward definite goals from which progress can be measured on a scale, such as daily workouts to achieve an optimal body mass index. The second category refers to such chain tales as "The Flying Ship" or "Tsarevitch Ivan, the Firebird and the Gray Wolf." In this kind of tale, the main character may have a clear goal, but the piling up of events seems to be in an arbitrary order.

Yet, all the events piled up together yield the final success. This category is the least character-driven. It provides analogous stories for mapping with goal-directed routines wherein completing a list of given tasks is of high priority. Examples include people attending a class or a training program that grants a certificate of attendance—the goal is to be present repeatedly. The third category includes chain tales like "The Stonecutter" in which the main characters' goals can be indefinite, but their intentions cause each event to be followed by the next "bigger" one. Hence, the gradual growth of events determines the order, but the end event is uncertain. The character-driven tendency of this category is medium. These tales can be good matches for routines that are driven by long-term visions or intangible goals whose endpoints are not well-defined, such as trading used goods or recycling plastic for sustainability, or regularly investing for long-term growth of wealth.

The above categorization of chain tales for different kinds of goal-directed routines examines the character-driven inclination in plot-driven narratives, in terms of the main characters' ultimate goals and their intentions toward each event. The two variables, namely, clear or unclear goals as well as intended or unintended event sequences, span two axes and form a matrix (Figure 5.1). The first category, most character-driven, sees clear goals and intended event sequences. In a sequence, successive events grow in size, and their order is fixed. The second category, least

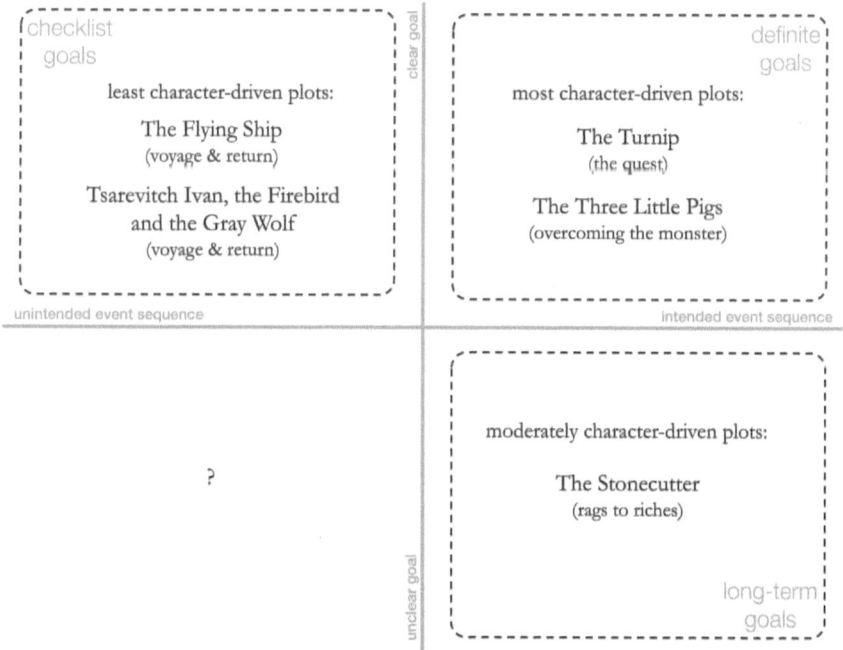

FIGURE 5.1 Categorization of chain tales for different kinds of goal-directed routines according to the character-driven inclination in the plot.

character-driven, refers to those with clear goals but unintended event sequences. Event order can be flexible. Unclear goals but intended event sequences belong to the third category, moderately character-driven, where the event order is fixed but the end event is not pre-determined. Finally, unclear goals and unintended event sequences form a very peculiar situation that is unusual in folktales. I shall discuss it in later sections.

Plot Types of Chain Tales for Routines

In a chain tale, characters' goals or intended actions, to a certain extent, affect the chain creation, that is, the plot. As E.M. Forster (1993) puts it, the plot emphasizes the links (mostly causal) between events. Many writers and theorists, building on huge volumes of varied human creative works, have proposed different narrative structure frameworks, including Propp (1968) on folktales from the Afanas'ev's collection, as well as Joseph Campbell (2004) on mythological narratives across different cultures. Both Propp's and Campbell's models delineate the commonly called "hero's journey" into a list of elemental events, which are organized in some specific order and can be selectively included in a tale. Between Propp's 31 "functions" and Campbell's 17 "stages," common ideas exist, for example, "receipt of a magical agent" in the former and "supernatural aid" in the latter. Although both Propp's and Campbell's models are influential in folklore studies, they are likely to include components that are unnecessary for a chain tale, whose plot is typically a chain of similar events that repeat the same action.

Instead of an all-inclusive framework, Christopher Booker (2004) in his study of classical novels as well as movies of the twentieth century, outlines seven basic plots that are commonly used. Among the seven basic types, four of them can be insightfully compared with the aforementioned three categories of chain tales. They are "the quest," "overcoming the monster," "rags to riches," and "voyage and return." Each of Booker's basic plots can be succinctly delineated in three parts. "The quest" starts with (1) someone setting out on a journey, followed by (2) overcoming obstacles, and finally (3) goals reached. Examples include the classic novel *The Lord of the Rings* and the Hollywood movie *Raiders of the Lost Ark*. "Overcoming the monster" is made up of (1) someone becoming aware of threats, (2) struggling, and finally (3) dark force overthrown. Movie series like *James Bond* and *Harry Potter* are prominent examples. "Rags to riches" typically tells that (1) the unrecognized acquires power, but (2) loses it, and finally (3) discovers new strength and succeeds. Common examples include the stories of *Cinderella* or *Aladdin*. "Voyage and return" is about someone (1) being precipitated in a strange world, (2) facing challenges and managing to pass or escape, and finally (3) returning with experience. The ancient Greek epic Odyssey is a classical example. Many others include Hollywood movies like *Back to the Future* and *The Lion King*. Booker's three-part summary of each basic plot seems to match the exposition, rising action, and falling action in Freytag's pyramid. The first part sets the stage, with a "journey" ahead or a strange world taking over. The second part commonly involves struggles. The third part brings success. These elements can be commonly found in chain tales, and the four basic plots singled out above can be used to label them. More importantly, the basic plots provide useful vocabularies for delineating chain tales into formal descriptions that facilitate analogical mapping and blending with corresponding routines.

In the first category of chain tales, "The Turnip" can be seen as "the quest" on a metaphorical journey. The peasant family's journey is to pull out the gigantic turnip from the ground, involving physical resistance as an obstacle, and they get what they want in the end. "The Three Little Pigs" obviously belongs to "overcoming the monster." The pigs are aware of threats from the wolf; they struggle to build more durable houses, and finally kill the wolf. In both "the quest" and "overcoming the monster," the protagonist has a clear goal to acquire an item (e.g., a gigantic turnip) or to defeat a villain (a malicious wolf), respectively. The event chain in either plot is a series of struggles involving increasingly more physical exertion or work, which is intended by the protagonist. In short, chain tales of plot type "the quest" or "overcoming the monster" can be put in the first category, which matches routines directed toward definite goals. The protagonist's work done can usually be quantified (e.g., the proportion of the turnip pulled out, the wolf's blowing force that a house withstands) and used to represent how far one has come along in the journey, as well as how far the definite goal is.

The second category of chain tales includes "The Flying Ship" and "Tsarevitch Ivan, the Firebird and the Gray Wolf," under the plot type of "voyage and return." The protagonist embarks on a journey to a strange land to complete missions with the hope of returning home. The simpleminded son in "The Flying Ship" hopes to acquire a flying ship and marry a princess, while the prince Ivan in "Tsarevich Ivan" is sent by his father to catch a firebird. Both protagonists start with a mission to accomplish. On their journeys, they encounter unexpected events, are given new missions, get help from companions, and finally make their ways to return home. Starting with clear goals but swamped by unintended events or challenges as the story unfolds, the plot type "voyage and return" matches the second category of chain tales, which can be applied to goal-directed routines entailing completion of a list of tasks within a period of time. An implicit checklist was prepared for the protagonist (e.g., different companions boarding the flying ship, varied tasks given by different tsars), and completing the list of tasks item by item indicates the progress a person has made in the routine.

The plot type of "The Stonecutter" or "The Fisherman and His Wife" in the third category is categorized under "rags to riches." The protagonist, an ordinary stonecutter or fisherman at first, makes increasingly greedier wishes that are granted by supernatural beings. The stonecutter finally reverts to his original self and realizes his unique talent. The fisherman and his wife lose everything but learn a lesson. In "rags to riches," the protagonist does not have a clear goal in the beginning. Acquiring power unexpectedly, one intends to make changes incrementally.

"Rags to riches" meets the conditions of the third category of chain tales, which works for routines in pursuit of long-term, indefinite goals. The protagonist experiences continual personal growth and it can provide a sense of achievement for a person engaged in the routine, although the distance to the endpoint is still not determined.

Case Study of Chain Tales

In the following sections, exemplary chain tales of each category are presented and discussed. Chain of events of each tale will be dissected in terms of composition and growth. The composition is typically "a series of" events that repeat a kind of action. The growth is about "increasingly" enlargement in form or treatment. Each tale is then categorized by its chain creation trait, including whether the characters' goals are clear and whether each recurring event is intended by the characters. Then I will talk about the plot type and synopsis, which suggest vocabularies for formal description of relations as presented in Chapter 4. The recurring events in the chain are similarly formalized for purposes of analogical mapping with corresponding representative routines. Lastly, new relations are suggested to be imported from external knowledge frames for blending with the routines.

The Turnip

Chain composition: a series of physical exertions
   Growth: increasingly more people and physical strength
   Goal: characters sharing the same clear goal, that is, to pull out the turnip
   Event sequence: all physical exertions intended by the characters
   Category: First category (definite goal)
   Plot type: The quest. The characters, starting with the grandfather, set out on a metaphorical journey to acquire the gigantic turnip. The physical resistance that keeps the turnip under the ground was the obstacle standing in their way. Finally, they reach their goal and get the turnip.
   Synopsis: A grandfather plants a turnip, which grows gigantic. He tries hard to pull it out but his efforts are in vain. He asks the grandmother to join, but the two still cannot pull the turnip out. They then ask their granddaughter to join, but the attempt is futile. They then get a dog to join, but the action remains unfruitful. A cat then joins the crew, but the turnip sits still in the soil. They then ask a mouse to join, and finally the six of them successfully pull the gigantic turnip out.
   Formal representation for mapping: Table 5.1
   Suggested routines: The tale can be mapped with goal-directed routines performed for definite, measurable goals, such as daily workouts to build muscles or lose weight. A person might set a daily weight training plan involving multiple pulling exercises, including barbell or dumbbell row, from Monday to Saturday. Imagine that after a workout on Monday, the person virtually sees the grandfather manages to pull only 15% of the turnip out of the soil. After a workout on Tuesday,

TABLE 5.1 The formal representation of events in "The Turnip" for mapping with routines

Iterative processes	Successive outcomes
JOURNEY (pulling together)	
OBSTACLE (physical resistance)	
GOAL (turnip)	
PULL (turnip, grandpa)	OUT (turnip, 15%)
ASK (grandpa, grandma)	
PULL (turnip, grandpa, grandma)	OUT (turnip, 30%)
ASK (grandma, granddaughter)	
PULL (turnip, grandpa, grandma, granddaughter)	OUT (turnip, 45%)
...	...

the person virtually sees the grandfather and the grandmother pulling 30% of it out. On Wednesday, joined by the granddaughter, the crew gets 45% of the turnip out, and so forth. On Saturday, the turnip completely comes out, and the person can take a break on Sunday.

Imported relations for blending: In the tale, after each unsuccessful attempt, the grandfather repeatedly asks one more family member for help. In reality, after each daily workout, the person can rest and let their muscles recover. They can come back the next day with more strength and stamina. Both the grandfather's work on the farm, including pulling up a turnip, and the person's workout routine, including the pulling exercises, can take place on a daily basis. To blend the tale with the workout routine, a knowledge frame external to both the tale and the routine is required to allow co-existence of components from both sides. The grandfather "commutes" daily from home to his farm. The weight lifter goes to the gym every day and might see the grandfather on the farm pulling a turnip, provided that the farm is close to the gym. While the gym is an urban locale, the recent practice of "urban farming" found in some cities renders possible the grandfather's farm present near the gym. The knowledge frame [urban farming] (the square brackets denote a frame) is invoked, a spatial relation NEAR (farm, gym) is imported, and a logical relation SEE (person, turnip), which means the weight lifter sees the turnip in the farm, can be inferred. After a workout, the person virtually sees the turnip being 15% out. The grandfather also goes back home to rest, and he asks a family member to join the next day. On the next day, the person goes to the gym again and virtually sees the grandfather and the grandmother pulling a turnip together. The turnip then is pulled 30% out. Whenever they work out, the person sees the grandfather and his family members pulling a turnip, probably on a wall-mounted digital display that simulates views of a farm next door in a window. This mixed-reality effect ties the workout routine with the fictional outcome of the pulling act. The person can imagine a third story wherein the farmer family works as hard as they do every day throughout the week, and share a fruitful reward on weekends. Table 5.2 summarizes the formal representation of the blending processes.

TABLE 5.2 The formal representation of the blends between the workout routine and "The Turnip"

Workouts	Selective projection and pattern completion	The Turnip
	[urban farming]: NEAR (farm, gym) SEE (person, turnip) PULL (turnip, grandpa)	← PULL (turnip, grandpa)
PULL (person, Day1)	→ PULL (person, Day1) OUT (turnip, 15%) [urban farming]: NEAR (farm, gym) SEE (person, turnip)	← OUT (turnip, 15%) ASK (grandpa, grandma)
PULL (person, Day2)	PULL (turnip, grandpa, grandma) → PULL (person, Day2) OUT (turnip, 30%) [urban farming]: NEAR (farm, gym) SEE (person, turnip)	← PULL (turnip, grandpa, grandma) ← OUT (turnip, 30%) ASK (grandma, granddaughter)
PULL (person, Day3)	PULL (turnip, grandpa, grandma, granddaughter) → PULL (person, Day3) OUT (turnip, 45%)	← PULL (turnip, grandpa, grandma, granddaughter) ← OUT (turnip, 45%)
…	…	…

The Three Little Pigs

Chain composition: a series of construction work by the pigs and attacks by the wolf

Growth: houses made of increasingly more resilient materials

Goal: characters share the same clear goal: to survive the wolf's attacks

Event sequence: all house building attempts intended by the characters

Category: First category (definite goal)

Plot type: Overcoming the monster. The characters are aware of the threat from the wolf. They build different houses to withstand the wolf's attacks but most of them fall apart. Finally, the third house survives, and the wolf dies.

Synopsis: Three little pigs live in a village. The first pig builds a house with straws, which is blown down by a malicious wolf. The second pig builds a house with sticks, which is also blown down by the wolf. The third pig builds a house with bricks, which is able to resist the wolf's blows. The wolf finally tries to climb down the chimney to catch the pigs. The third pig then boils a pot of water at the fire place, in which the wolf falls and dies.

Formal representation for mapping: Table 5.3

Suggested routines: The tale matches goal-directed routines performed for definite goals, like taking medications or supplements for a period of time, probably

TABLE 5.3 The formal representation of events in "The Three Little Pigs" for mapping with routines

Iterative processes	Successive outcomes
THREAT (wolf)	
STRUGGLE (pigs, blows)	
GOAL (wolf removed)	
BUILD (first pig, first house, straws)	BLOWN (straws, wolf)
BUILD (second pig, second house, sticks)	BLOWN (sticks, wolf)
BUILD (third pig, third house, bricks)	BLOWN (bricks, wolf)

with regular physical exercises, in order to strengthen immunity. Another similar challenging routine can be a parent making their child take medicine regularly. People might be prescribed with medications due to sickness or under certain health condition, or one might intentionally take supplements to boost health. A person starts taking a pill regularly and virtually sees a house built with straws being blown down by a wolf. The house fails to withstand the wolf's blow even when the wolf is just using 50% of its strength. In the second week, one maintains the routines and virtually sees another house built with sticks. Again, it does not survive the wolf's attack when the wolf blows it in full strength. In the third week, the person gets to see a house built with bricks finally able to withstand the wolf's blows as long as they still engage in the pill-taking routines.

Imported relations for blending: In the tale, the pigs successively build more resilient houses to withstand the wolf's attacks. In reality, the person regularly takes medications or supplements to strengthen immunity and fight disease. To connect the tale with the pill-taking plus exercise routines, an external knowledge frame [neighborhood] can be considered. From the knowledge frame, a spatial relation NEAR (home, pig's house), which means the pigs' house is near the person's home, can be imported. Virtually seeing that the pig's house nearby is blown by the wolf, probably on a wall-mounted digital display simulating views through a window, the person can imagine the wolf's blows resulting in the recent windy weather. Another logical relation that follows is CAUSE (BLOWN (pig's house, wolf), SICK (person)). The person might have got a cold (i.e., a viral infection) due to the windy weather. Running the blend, the person can imagine a third story wherein they are a neighbor of the pigs. Together they face the gusts and defend themselves together. When the person takes the pills in the first week, they virtually see the first house built with straws, but it is too weak to withstand the wolf's blow, like the person's immune system. In the second week, the pigs build the house with sticks, but it is still not able to withstand the wolf's attack. In the third week, the house of bricks is finally strong enough to survive the wolf's ambush. The person also gets better and sees the wolf climbing down the chimney of the pigs' house. They can tell the pigs, and the threats are finally gone. Table 5.4 summarizes the formal representation of the blending processes.

TABLE 5.4 The formal representation of the blends between taking medications and "The Three Little Pigs"

Taking medications		Selective projection and pattern completion	The Three Little Pigs
		BUILD (pig, house of straws) [neighborhood]: NEAR (home, house of straws)	← BUILD (pig, house of straws)
SICK (person)	→	CAUSE (BLOWN (house of straws, wolf), SICK (person)) CAUSE (BLOWN (house of straws, wolf), DOWN (house of straws))	← CAUSE (BLOWN (house of straws, wolf), DOWN (house of straws))
TAKE (person, medications, Week1)	→	TAKE (person, medications, Week1)	
		BUILD (pig, house of sticks) [neighborhood]: NEAR (home, house of sticks)	← BUILD (pig, house of sticks)
SICK (person)	→	CAUSE (BLOWN (house of sticks, wolf), SICK (person)) CAUSE (BLOWN (house of sticks, wolf), DOWN (house of sticks))	← CAUSE (BLOWN (house of sticks, wolf), DOWN (house of sticks))
TAKE (person, medications, Week2)	→	TAKE (person, medications, Week2)	
		BUILD (pig, house of bricks) [neighborhood]: NEAR (home, house of bricks)	← BUILD (pig, house of bricks)
		BLOWN (house of bricks, wolf)	← BLOWN (house of bricks, wolf)
SICK (person)	→	CAUSE (BLOWN (house of sticks, wolf), SICK (person))	
TAKE (person, medications, Week3)	→	TAKE (person, medications, Week3)	
RECEOVER (person)	→	STAND (house of bricks) RECEOVER (person)	← STAND (house of bricks)
		CLIMB (wolf, chimney)	← CLIMB (wolf, chimney)
		TELL (person, pigs)	
		BOIL (pigs, water, wolf)	← BOIL (pigs, water, wolf)

The Flying Ship

Chain composition: a series of unexpected encounters
   Growth: increasingly more people on board
   Goal: characters have a clear goal.
   Event sequence: encounters in the chain are not intended by the characters
   Category: Second category (checklist goal)

Plot type: Voyage and return. The main character comes across a flying ship and encounters people with varied, peculiar powers. They are given different deadly missions by the king. Their peculiar powers save them.

Synopsis: The third son of an old peasant couple is a simpleton, but is determined to get a flying ship in order to marry the princess. Setting out on his journey with no plan, he comes across a mysterious old man who helps him build a ship that flies. He then comes across a man with incredible hearing capability (the listener) and invites him to board the ship. Next, he invites a man who can run incredibly fast (the runner) to join him on board. Then, someone with incredible eyesight (the archer), incredible appetite (the gobbler), incredible thirst (the guzzler), a peculiar power to turn straw into snow (the snowmaker), and an ability to turn magic wood into solders (the wood-carrier) are all invited to board the flying ship one by one. Finally, their flying ship arrives at the king's castle so that the simpleton can marry the king's daughter. To avoid giving away his daughter to these peasants, the king wickedly gives them impossible and deadly missions. First, they must bring life-giving water; the tasks are then accomplished by the listener, the runner, and the archer. Second, they must eat an incredible amount of food; the task is then accomplished by the gobbler. Third, they must drink an incredible amount of wine, the mission is accomplished by the guzzler. Fourth, the simpleton must bath in scorching hot water, which is then cooled down by the snowmaker. The king finally sends an army to kill the simpleton, but the wood-carrier turns magic wood into soldiers to defend. They flee aboard successfully with the princess and return home.

Formal representation for mapping: Table 5.5

Suggested routines: The tale, like some voyage and return stories, is about the protagonist building a peculiar team in a strange realm. The composition of the team is more important than the order in which members join the team. When it is mapped with goal-directed routines, the team-building criterion matches well the completion of a checklist. Let's suppose a person who has enrolled in a three-month course needs to attend the class once a week. When they commute to the community college for the first class, they virtually meet and recruit the listener, probably opening an instant messaging chat group called "Flying Ship" and seeing the

TABLE 5.5 The formal representation of events in "The Flying Ship" for mapping with routines

Iterative processes	Successive outcomes
GOAL (marry princess)	
CHALLENGE (king)	
INVITE (fool, listener)	ON BOARD (fool, listener)
INVITE (fool, runner)	ON BOARD (fool, listener, runner)
INVITE (fool, archer)	ON BOARD (fool, listener, runner, archer)
...	...

listener join the group. In the second class, the person virtually recruits the runner, who also joins the chat group. Every class corresponds to one unique team member joining the "Flying Ship" chat group. After three months, the person has built a full team consisting of 12 different "talented" members, ready for taking the final examination or any future challenges with the talents acquired.

Imported relations for blending: In the tale, the protagonist invites strangers with different talents to board the flying ship and builds a full team. In real life, the person enrolling in a course learns something different in each class. Hence, every class corresponds to one unique member in the tale. To blend real-world class learning with the special skill of each member in the tale, an external knowledge frame [team project] may be invoked. New logical relations that can be imported from the knowledge frame include TALENT (new member, new skill), PEER (person, new member), and LEARN FROM (person, new member). Whenever a new member boards the flying ship, he becomes part of the team. The simpleton is able to learn some special skills from his peers. In each real-world class, the person might team up with a classmate for a group project. The person might learn something from their peers too. The new logical relations bind together the act of being peers (either on the same ship or in the same group) with the act of learning. Running the blend, the person can imagine a third story of themselves boarding the flying ship with talented classmates and accumulating different skills for the final examination or future challenges. Table 5.6 summarizes the formal representation of the blending processes.

Tsarevitch Ivan, the Firebird, and the Gray Wolf

Chain composition: a series of unexpected tasks

    Growth: increasingly more unfinished tasks piled up

    Goal: Characters have clear goals in the beginning

    Event sequence: All tasks and their outcomes in the chain are not expected by the characters.

    Category: Second category (checklist goal)

    Plot type: Voyage and return. The main character is sent to accomplish a task and comes across the gray wolf with peculiar power. Every time before he finishes a task, the character is sent to accomplish another. The gray wolf finally solves all the problems.

    Synopsis: The king's youngest son, Ivan, wants to help his father catch the firebird that has stolen the golden apples. He comes across a gray wolf with special power. He rides on the gray wolf and tries to steal the firebird from another king but he gets caught. To have the firebird, he has to catch a horse with golden mane in exchange. He rides the gray wolf again and tries to steal the horse with golden mane from yet another king but he gets caught again. To have the horse with golden mane, he has to catch Yelanda the Beauty. He rides the gray wolf and tries to catch Yelanda but ends up falling in love with her. The gray wolf transforms itself into

TABLE 5.6 The formal representation of the blends between attending classes and "The Flying Ship"

Attending classes	Selective projection and pattern completion	The Flying Ship
ATTEND (person, class, Week1)	→ ATTEND (person, class, Week1)	
	TALENT (listener, skill1) [team project]:	← TALENT (listener, can hear from a far distance)
	PEER (person, listener)	INVITE (simpleton, listener)
	LEARN FROM (person, listener)	
LEARN (person, skill1)	→ LEARN (person, skill1)	ON BOARD (simpleton, listener)
ATTEND (person, class, Week2)	→ ATTEND (person, class, Week2)	
	TALENT (runner, skill2) [team project]:	← TALENT (runner, can run a long distance)
	PEER (person, runner)	INVITE (simpleton, runner)
	LEARN FROM (person, runner)	ON BOARD (simpleton, listener, runner)
LEARN (person, skill2)	→ LEARN (person, skill2)	
ATTEND (person, class, Week3)	→ ATTEND (person, class, Week3)	
	TALENT (archer, skill3) [team project]:	← TALENT (archer, can shot a far target)
	PEER (person, archer)	INVITE (simpleton, archer)
	LEARN FROM (person, archer)	ON BOARD (simpleton, listener, runner, archer)
LEARN (person, skill3)	→ LEARN (person, skill3)	
...	...	...

Yelanda and helps all flee. The gray wolf then turns itself into the horse with golden mane, followed by the firebird, and again helps all flee. Finally, the youngest son safely returns home with the firebird, the horse with golden mane, and Yelanda.

Formal representation for mapping: Table 5.7

Suggested routines: Unlike "The Flying Ship," which is about piling up of team members, this tale is about piling up of unfinished tasks. It is particularly applicable to routines like piecework, such as delivering online shopping orders or food, handyman jobs, or meetings with clients. Let's suppose a person commutes to do piecework, which typically appears on demand. On the first day, they virtually see Ivan riding on a gray wolf trying to catch the firebird, probably through an augmented-reality (AR) headset or a mobile AR app. At the end of the first day, the person virtually sees the firebird caught. When the person commutes to do

TABLE 5.7 The formal representation of events in "Tsarevitch Ivan, the Firebird and the Gray Wolf" for mapping with routines

Iterative processes	Successive outcomes
RIDE (Ivan, Gray Wolf)	
CATCH (Ivan, Firebird)	HOLD (first king, Firebird); ASK (first king, Ivan, horse)
CATCH (Ivan, horse)	HOLD (second king, horse); ASK (second king, Ivan, Yelanda)
CATCH (Ivan, Yelanda)	LOVE (Ivan, Yelanda)
TRANSFORM (Gray Wolf, Yelanda)	FLED (Ivan, Yelanda)
TRANSFORM (Gray Wolf, horse)	FLED (Ivan, Yelanda, horse)
TRANSFORM (Gray Wolf, Firebird)	FLED (Ivan, Yelanda, horse, Firebird)

piecework the second day, they see Ivan on the gray wolf again trying to catch the horse with golden mane. When the person performs piecework on the third day, they see Ivan on the gray wolf trying to catch Yelanda, and so forth. On the last day before the holiday, the person accomplishes all the duties on the checklist and sees the gray wolf help Ivan flee back home with all the rewards.

Imported relations for blending: In the tale, Ivan rides the gray wolf to undertake one given task. Before finishing it, he was caught and given another task. He might set out on the next day and ride the gray wolf again to undertake the second task. Unfinished tasks accumulated each day. In reality, the person commutes to do piecework day after day. The person might name each day's piecework with a precious creature or the Beauty, such as the firebird or the horse with golden mane. On the last working day before the weekend, the person successfully finishes all the piecework and feels fruitful on the return commute. In short, both Ivan's undertaking and the person's piecework can be named the same in a knowledge frame [mission titles]. From the knowledge frame, a new relation can be imported for blending is NAME (mission, title), which assigns a title to a mission. NAME (DAY 1, Firebird) binds together the person's first-day piecework and Ivan's first undertaking. NAME (DAY 2, horse with golden mane) relates one's second-day piecework with Ivan's second undertaking, and so forth. Another external knowledge frame to be invoked can be [named trains], which is the practice of giving a passenger train on a regular route a special name. The person can name the train he or she rides as "gray wolf." The person can imagine a third story wherein they see the commute vehicle as the gray wolf with magical power helping them accomplish the piecework every day. Table 5.8 summarizes the formal representation of the blending processes.

The Stonecutter

Chain composition: a series of requests
    Growth: increasing greed for power
    Goal: characters do not have clear and definite goals in the beginning
    Event sequence: Every wish made in the chain is intended by the main character.

TABLE 5.8 The formal representation of the blends between commuting and "Tsarevitch Ivan, the Firebird and the Gray Wolf"

Commuting to do piecework	Selective projection and pattern completion	Tsarevitch Ivan, the Firebird and the Gray Wolf
WORK (person, Day1)	→ WORK (person, Day1) [mission titles] NAME (Day 1, Firebird) CATCH (person, Firebird)	← CATCH (Ivan, Firebird)
RIDE (person, train, Day1)	→ RIDE (person, train, Day1) [named trains] NAME (train, Gray Wolf) RIDE (person, Gray Wolf) HOLD (first king, Firebird)	← RIDE (Ivan, Gray Wolf) ← HOLD (first king, Firebird)
	ASK (first king, person, horse)	← ASK (first king, Ivan, horse)
RECEIVE (person, next work)	→ RECEIVE (person, next work)	
WORK (person, Day2)	→ WORK (person, Day2) [mission titles] NAME (Day 2, horse) CATCH (person, horse)	← CATCH (Ivan, horse)
RIDE (person, train, Day2)	→ RIDE (person, train, Day2) [named trains] NAME (train, Gray Wolf) RIDE (person, Gray Wolf) HOLD (second king, horse)	← RIDE (Ivan, Gray Wolf) ← HOLD (second king, horse)
	ASK (second king, Ivan, Yelanda) RECEIVE (person, new job)	← ASK (second king, Ivan, Yelanda)
RECEIVE (person, next work)	→	
WORK (person, Day3)	→ WORK (person, Day3) [mission titles] NAME (Day 3, Yelanda) CATCH (person, Yelanda)	← CATCH (Ivan, Yelanda)
RIDE (person, train, Day3)	→ RIDE (person, train, Day3) [named trains] NAME (train, Gray Wolf) RIDE (person, Gray Wolf)	← RIDE (Ivan, Gray Wolf) LOVE (Ivan, Yelanda)
	TRANSFORM (Gray Wolf, Yelanda) TRANSFORM (Gray Wolf, horse) TRANSFORM (Gray Wolf, Firebird) FLED (person, Yelanda) FLED (person, Yelanda, horse)	← TRANSFORM (Gray Wolf, Yelanda) ← TRANSFORM (Gray Wolf, horse) ← TRANSFORM (Gray Wolf, Firebird) ← FLED (Ivan, Yelanda)
	FLED (person, Yelanda, horse, Firebird)	← FLED (Ivan, Yelanda, horse) FLED (Ivan, Yelanda, horse, ← Firebird)

Category: Third category (long-term goal)

Plot type: Rags to riches. The character is an ordinary stonecutter and his wishes are granted by supernatural beings. He wishes to become more and more powerful, but finally he just wants to be a stonecutter again with a lesson learnt.

Synopsis: A stonecutter comes across a mountain spirit who promises to make his wishes come true. At first, he wishes to be a rich man and he gets it. He then thinks a prince is more powerful than a rich man. So he wishes to be a prince and he gets it. Later on, he finds that the sun is even more powerful, and so he makes a wish and becomes the sun. He then sees a cloud blocking the sun, and wishes to be a cloud. After that, he sees a cloud pour heavy rain to destroy everything but the rock. So he makes a wish and becomes a rock. Finally, he realizes that a stonecutter can break the rock with tools. So he wishes to be a stonecutter again.

Formal representation for mapping: Table 5.9

Suggested routines: The tale can be considered for mapping with those routines driven by long-term, distant, and intangible goals, like bartering used items for new goods or recycling items such as plastic bottles or aluminum cans. A person who often trades used clothes or accessories like handbags for new goods might do it because they believe in a circular economy. One can continue to enjoy novel product experiences at lower prices, meanwhile giving used items a new lease of life with a new owner. On the other hand, to those with a habit of categorizing and recycling used containers and the like, the routine acts would benefit society and nature in the long run. The economic or sustainable goals can be intangible. Imagine someone who trades a used item for a new good. For the first time, they virtually see the stonecutter turning into a rich man, probably on a mobile AR app. When trading the second time, they virtually see the rich man turning into a prince. The third time around the prince becomes the sun, and so forth. When the person sees him turn back into a stonecutter, the meaning of cycle resonates in both the tale and real life. One might think that it is time to recycle.

Imported relations for blending: In the tale, the protagonist consecutively wishes to become another more powerful being. In reality, the person, who trades used items for new goods, is engaged in an ecosystem of production and consumption. After consuming a good, the person might recycle it. This ecosystem can be comparable to the mountain spirit in the tale, who realizes the stonecutter's wishes,

TABLE 5.9 The formal representation of events in "The Stonecutter" for mapping with routines

Iterative processes	*Successive outcomes*
POWER (stonecutter, wish)	
WISH (stonecutter, rich-man)	BECOME (stonecutter, rich-man)
WISH (rich-man, prince)	BECOME (rich-man, prince)
WISH (prince, sun)	BECOME (prince, sun)
...	...

transforming him into one form, then another, and finally back to his original self. In fact, an ecosystem can be so huge that it includes social, economic, environmental, cultural, and even mythological components, and they are all connected in immensurable and imperceptible ways. The external knowledge frame that can be brought in for blending is the phenomenon of chain reaction in a complex eco-system, which can be hastily summarized as changes in one component resulting in changes elsewhere. When the stonecutter makes a new wish, he also gives up his current power in exchange for new power. When the person trades a used item, they actually give up their current right of possession in exchange of possessing a new item. The new relation EXCHANGE (current item, new item) encompasses both the act of making a wish in the tale and the act of trading in reality. Another new relation PROPOGATE () then connects EXCHANGE () in trading items with EXCHANGE () in the tale. When the person trades in a used item, the action triggers an effect, which propagates in the complex eco-system and distantly leads to the stonecutter's wish for new power being granted. The stonecutter's sacrificed old power in turn distantly links to the new good that the person gets. Seeing this distant "butterfly effect," the person can imagine a third story about the "entanglement" between one's act of trading in or recycling with a distant stonecutter's wishes. At the end of the tale, when the stonecutter-turned rock wishes to be a stonecutter again, he also returned all the powers he once had to the mountain spirit. The person trading in might consider returning everything they once own to the eco-system as well, that is, recycling an item after consuming it and hoping that it will be used again to produce the item in its original state. Table 5.10 summarizes the formal representation of the blending processes.

Theme-Based Narratives for Maintenance Routines

The categorization of chain tales for different kinds of routines examines two independent variables, forming the matrix of four quadrants. The three quadrants correspond to three categories used to match an array of goal-directed routines that vary in their types of goals, including goals that can be well-defined on a scale of progress, goals consisting of a checklist to be completed, and long-term goals. In the fourth quadrant, where both variables are negative, characters' goals are not obvious, and the characters do not intend the recurring events to happen. In other words, characters do not play a key role in those events. They might fade into settings or even become implicit. Without characters' goals and intended actions, the reasons for or causes of recurring events also become obscure. The absence of causal links also weakens the plot. In this peculiar quadrant, a story is neither character-driven nor plot-driven.

If a story in this quadrant does exist, actions are still repeated in successive events in an implicit way, while varying in other parts. That is, the event sequence features variations over progression. This kind of story is suitable for mapping with maintenance routines, because people might want to see variety during the

TABLE 5.10 The formal representation of the blends between trading items and "The Stonecutter"

Trading used items	Selective projection and pattern completion	The Stonecutter
OWN (person, item1) →	OWN (person, item1)	
TRADE-IN (person, item1, new item2) →	TRADE-IN (person, item1, new item2) [chain reaction]:	
	WISH (stonecutter, rich-man)	← WISH (stonecutter, rich-man)
	PROPROGATE (EXCHANGE (item1, new item2), EXCHANGE (stonecutter, rich-man))	
	BECOME (stonecutter, rich-man)	← BECOME (stonecutter, rich-man)
OWN (person, item2) →	OWN (person, item2)	
TRADE-IN (person, item2, new item3) →	TRADE-IN (person, item2, new item3) [chain reaction]:	
	WISH (rich-man, prince)	← WISH (rich-man, prince)
	PROPROGATE (EXCHANGE (item2, new item3), EXCHANGE (rich-man, prince))	
	BECOME (rich-man, prince)	← BECOME (rich-man, prince)
OWN (person, item3) →	OWN (person, item3)	
TRADE-IN (person, item3, new item4) →	TRADE-IN (person, item3, new item4) [chain reaction]:	
	WISH (prince, sun)	← WISH (prince, sun)
	PROPROGATE (EXCHANGE (item3, new item4), EXCHANGE (prince, sun))	
	BECOME (prince, sun)	← BECOME (prince, sun)
...	...	...

repetitive and boring processes rather than anticipating progress. This quadrant suggests rethinking the general definition of "story" as a sequence of causally *or* thematically related events. When the causal relation between successive events is not clear, the narrative structure can be hinged on other relations like chronological (orderly events arranged in time sequence), procedural (orderly events prescribed by established systems), spatial (events organized in multiple dimensions), or even categorical (events organized in categories). For example, Monet's *Haystacks* series (1890–1891) can be seen as a pictorial narrative of chronologically and categorically related events. All depicted events repeat the same motif and similar settings, that is, haystacks in the field. They show no apparent actions or characters but instead results of preceding actions (e.g., stacking up of harvested barley) by implicit characters (e.g., a farmer). The characters' goals are not presented, and it is unknown whether the actions are intended choices or not. The emphasis falls on, not characters or actions, but instead varied outcomes. This forms a special category, what I call "theme-based narratives." The pictures vary in lighting and environmental details, which inform the events happening at different times of day and different seasons. A chronological order can be sorted out. Hence, the *Haystacks* series presents a narrative.

Monet is one of the most renowned Western artists who loved to repeatedly explore the same subject and created what are known today as "series paintings." Yet, he was by no means the first artist creating and publishing series works. In the Edo period of Japan (1603–1868), with advances in woodblock printing and emergence of varied entertainment businesses, quite a few masters in ukiyo-e (a school of Japanese art) created and published "series prints," not only sparking high demands at that time but also remaining well-regarded today. Prominent examples include Katsushika Hokusai's *36 Views of Mount Fuji* (c. 1830–1832) and Utagawa Hiroshige's *100 Famous Views of Edo* (1856–1858). Like Monet, Hokusai addressed the same subject, Mount Fuji, in the series, presenting it from 36 vintage points at different locations around the mountain. The series can be seen as a pictorial narrative of spatially related events about gazing and appreciating the iconic mountain. Hiroshige, on the other hand, dealt with another theme, the city of Edo, in terms of people, streets, activities, buildings, and the like. The series consists of more than 100 prints covering four seasons, which can be seen as a pictorial narrative of categorically and semi-chronologically related events that celebrate the vibrant city. These series show and tell stories in which the recurrent themes are not characters but settings.

Japanese Woodblock Prints

While folktales verbally describe actions, characters, and settings that are able to evoke images, as seen in many picture books or miniatures on lacquer boxes,[9] the visual arts such as paintings and woodblock prints can directly project images of story components. Woodblock, as an art form, has started to acquire popularity in the Edo period of Japan, alongside the rise of the unique style called "ukiyo-e," which literally means "pictures of the floating world." It is said that the term "ukiyo" originally referred to a Buddhist expression about the sorrowful world. As Japanese woodblock became popular for depicting and promoting entertainment businesses such as the brothel or the theater, the term "uki" originally meaning "sorrow" was replaced by another similarly sounding kanji meaning "floating." The sorrow world transformed into the floating world, which poetically describes the transient and fleeting pleasures in these worldly entertainments. With "e" meaning "picture," ukiyo-e now commonly refers to Japanese pictorial art, particularly woodblock prints, from the Edo period.[10]

The initial woodblock prints in Japan were only black and white and mostly found in illustrated books. The polychrome printing technique, with multiple blocks each registering different color, started to appear around 1765. Meanwhile, creative print designers relentlessly pursued innovative works that could stand out from the illustrated books as individual pieces of art. Okumura Masanobu (c. 1684–1764) incorporated Western-style linear perspective as "uki-e," that means perspective picture, famously depicting such scenes as kabuki theaters (a traditional form of Japanese drama), which were favorites of the increasing population who could afford entertainments and purchase prints for leisure. Suzuki Harunobu (1724–1770) was known to be the first print designer harnessing the technology to produce full-color prints, called "nishiki-e." Kitagawa Utamaro (1753–1806) was famous for his portraits of women, while Tōshūsai Sharaku, though being active only for less than a year, left behind well-known works portraying many kabuki actors, who were comparable to today's Hollywood superstars.

The later phase of ukiyo-e shifted its subject matters from theatrical entertainments or celebrities to landscape and topics related to travel, owing to the imposition of the sankin kōtai (alternate attendance) system. As the city of Edo became new capital, the military government of the period, Tokugawa shogunate, required feudal lords, daimyō, to alternate residence between one's home domain and Edo. For the system to work, the famous five major roads were developed from Edo to other domains, including the old capital Kyoto. With the well-developed infrastructure, as well as the rise of the middle class, travel for leisure or pilgrimage became a fashionable activity during the late Edo period. Ukiyo-e prints of attractions or street views at different places were produced and popularly received as souvenirs for those who traveled or others who could not. Hokusai's *36 Views of Mount Fuji* displays different scenic views of the iconic mountain, which seem to promote attractions at different places in the region. Hiroshige's *100 Famous Views of Edo* explicitly presents images of different attractions in the city, many of which include iconic objects in the foreground, people in the street, and a spectacular view in the background. Another famous series work by Hiroshige, *53 Stages of the Tōkaidō*, seems to show and tell a travel story from Edo to Kyoto along the Tōkaidō, one of five major roads in the Edo period, capturing the landscapes, guesthouses, shops, and travelers on the road. It was followed by a sequel, *69 Stages of the Kisokaidō*, focusing on another road also connecting Edo and Kyoto. Each of these ukiyo-e print series visually presents a story made up of a sequence of consistent yet varied events, constituting a virtual journey back to the Edo period. The consistency in composition and presentation of the events with variations in the contents render a print series of this kind appropriate for mapping with maintenance routines. When practicing a routine to maintain a lifestyle grounded in a belief, such as walking stairs every day, mindfully eating every meal, or doing the laundry twice a week, a person would enjoy seeing some colors during the repetitive acts.

Stories for mapping with maintenance routines hence have the following characteristics: (1) the rising action features variations more prominently than progression; (2) characters can be implicit; and (3) successive events, which repeat some components but vary in others, may not be causally related. Ukiyo-e print series in the late Edo period, with each explicit theme on tourist attractions, street views, or landscapes in the floating world, project images of the kind of story for mapping with maintenance routines.

Case Study of Ukiyo-e Print Series

In the following sections, several well-known ukiyo-e print series that intriguingly match maintenance routines are presented and discussed. Each print series here is regarded as a pictorial narrative, and hence interpreted as a series of events first and foremost. Each print can be seen as an event. Although successive events repeat some story components, it is the variations among them that add color to the monotonous maintenance routines. How the prints in a series are consistent but varied is the second topic to be discussed. It is followed by the introduction of the collection in the series, with focus on the visual contents presented in the prints and the visual compositions (i.e., framing). The final parts, like the analyses of chain tales, cover the synopsis (as it is a story after all!), formal representation for mapping, suggested routines, and lastly new relations to be imported for blending.

36 Views of Mount Fuji (by Hokusai)

Events: Each print in the series presents a view of Mount Fuji taken from a place named in the title (with a few exceptions), which can be regarded as an event recorded there. The series forms a narrative of thematically related events.

Variation: The iconic mountain repeatedly appears in all prints, with different environments, people, activities, animals, seasons, and viewing angles, projecting images of varied events.

Collection: The whole series consists of the initial 36 pieces, plus the additional ten pieces responding to the contemporary positive reception of the original series. The publication of the initial series is believed to start with ten pieces printed mainly in Prussian blue and signed by 「前北斎為一筆」 (literally meaning "former Hokusai as one stroke"), followed by another 16 pieces printed with multiple colors and signed the same, and then ten more pieces signed by 「北斎為一筆」 (literally meaning "Hokusai as one stroke").[11] Each batch includes a few prints that are worth mentioning for analysis here (see Table 5.11).

Content: Some prints present mainly natural environments with barely any human activity in sight, like "Umezawa Hamlet-fields in Sagami Province." In some pieces, human activity can be recognized, such as barrel-making, house-repairing, sightseeing, fishing, farming, and woodcutting. Other prints cover people who are travelers or transporters. The view in the most famous print, titled "Through the Waves off

TABLE 5.11 Selection of prints from each batch of *36 Views of Mount Fuji* with the corresponding content and viewing angle

Print title	Content	Viewing angle
First batch:		
Hongan-ji Temple at Asakusa in Edo	kite, temple, rooftop	mid-altitude view
Umezawa Hamlet-fields in Sagami Province	Japanese cranes	high-altitude view
Kajikazawa in Kai Province	fishing, cliff, waves	low-altitude view
In the Totomi Mountains	woodcutting, smoke	low-altitude view
Lake Suwa in Shinano Province	lake, hut, boat	high-altitude view
Second batch:		
Snowy Morn at Koishikawa	snow, sightseeing, huts	high-altitude view
Lower Meguro	farming	mid-altitude view
Nihonbashi Bridge in Edo	river, houses, travelers	low-altitude view
The Bay of Noboto	torii (gateway of a shrine), transporters, residents	mid-altitude view
Hakone Lake in Sagami Province	lake	high-altitude view
Suruga-cho Street in Edo; the Mitsui Shop	house-repairing	mid-altitude view
Third batch:		
Fujimi-ga-hara in Owari Province	barrel-making	low-altitude view
Under Mannen Bridge at Fukagawa	bridge, river, boat	low-altitude view
Through the Waves off Kanagawa	waves, boats	mid-altitude view
Thunderstorm Beneath the Summit	lightning	high-altitude view
On a Fine Breezy Day	clear sky	high-altitude view

Kanagawa," was taken near the sea at Kanagawa, capturing waves and boats, in front of Mount Fuji. The view in the print titled "On a Fine Breezy Day" was taken from an unknown location, yet the weather and the red mountain implied that the season was autumn. As the title of the print "Thunderstorm Beneath the Summit" said, lightning appeared in the view. The outline of the summit in the print suggested that the view was taken from the west of Mount Fuji. Most other prints were named with a location, such as "Under Mannen Bridge at Fukagawa," "Nihonbashi Bridge in Edo," "Hakone Lake in in Sagami Province," and "Kajikazawa in Kai Province."

Framing: Some prints present a high-altitude view, that is, viewing Mount Fuji from a location at high altitude, say, from the top of another mountain, like "Lake Suwa in Shinano Province" and "Snowy Morn at Koishikawa." Other prints show a mid-altitude view, which is like taken from the top of a house, such as "Hongan-ji Temple at Asakusa in Edo" and "The Bay of Noboto." Lastly, some prints feature a low-altitude view like taken in the street, such as "Nihonbashi Bridge in Edo" and "Under Mannen Bridge at Fukagawa."

TABLE 5.12 The formal representation of events in *36 Views of Mount Fuji* for mapping with routines

Iterative processes	Successive outcomes
...	
TRAVEL (Asakusa)	
VIEWPOINT (mid-altitude, Fuji)	APPEAR (Fuji, kite, temple, rooftop)
...	
TRAVEL (Koishikawa)	
VIEWPOINT (high-altitude, Fuji)	APPEAR (Fuji, snow, sightseeing, huts)
...	
TRAVEL (Fukagawa)	
VIEWPOINT (low-altitude, Fuji)	APPEAR (Fuji, bridge, river, boat)
...	

Synopsis: The series can be interpreted as a pictorial narrative. An implicit and imaginary character traveled to different places to look at Mount Fuji, appreciating the beauty of nature, sometimes also reporting the people engaged in varied activities.

Formal representation for mapping: Table 5.12

Suggested routines: This print series consistently presents the views of Mount Fuji taken from varied places, which constitute a narrative featuring variations more than progression. As Mount Fuji is a fixed target, changes of the viewer's location result in different viewing angles. The variation thus lies not only in the content like the environment or human activity but also in the visual composition. The composition of each print reveals that the viewer's place is at a high (e.g., from another mountain) or low (e.g., on a street) altitude. Hence, the print series can be considered for mapping with maintenance routines that involve changing places, particularly of different heights. Taking stairs instead of an elevator in daily life is a good match. Consider a person regularly walking up the stairs to office or back home. One might not aim to lose weight but just believe that the habit is good for health and the environment (through saving electricity). When the person climbs two or three floors, they see a view of Mount Fuji from a low altitude, such as "Nihonbashi Bridge in Edo." After climbing a few more floors, they see a view from a mid-altitude, like "Hongan-ji Temple at Asakusa in Edo." If they consistently walk the stairs for a week, they get to enjoy the view from an even higher angle, like "Hakone Lake in Sagami Province" or "On a Fine Breezy Day."

Imported relations for blending: In the series, the imaginary viewer travels to different places and looks at Mount Fuji from different levels and angles. For those prints with a low-altitude view, the imaginary viewer just stays on the street level. For those prints with a mid-level view, the viewer walks up, like to the upper floor of a house, to earn the view. For those prints with a high-altitude view, they might need to climb to the top of a hill to see Mount Fuji. In real life, the person climbs stairs from a lower floor to a higher floor inside a building. If there are windows in the staircase,

TABLE 5.13 The formal representation of the blends between stair climbing and *36 Views of Mount Fuji*

Taking stairs	Selective projection and pattern completion	36 Views of Mt. Fuji
	[observation tower]:	
	DECK (5th Floor, mid-altitude)	TRAVEL (Asakusa)
CLIMB (5th → Floor)	CLIMB (5th Floor)	
	VIEWPOINT (mid-altitude)	← VIEWPOINT (mid-altitude)
	APPEAR (Fuji, kite, temple, rooftop)	← APPEAR (Fuji, kite, temple, rooftop)
	[observation tower]:	
CLIMB (9th → Floor)	DECK (9th Floor, high-altitude)	TRAVEL (Koishikawa)
	CLIMB (9th Floor)	
	VIEWPOINT (high-altitude)	← VIEWPOINT (high-altitude)
	APPEAR (Fuji, snow, sightseeing, huts)	← APPEAR (Fuji, snow, sightseeing, huts)
	[observation tower]:	
	DECK (3rd Floor, low-altitude)	TRAVEL (Fukagawa)
CLIMB (3rd → Floor)	CLIMB (3rd Floor)	
	VIEWPOINT (low-altitude)	← VIEWPOINT (low-altitude)
	APPEAR (Fuji, bridge, river, boat)	← APPEAR (Fuji, bridge, river, boat)
...	...	...

the person can see the views outside. When the person walks up to a floor that is high enough, one might see mountains from a distance. To blend real-world stair climbing with different views of Mount Fuji, the external frame [observation tower] can be invoked. It is a kind of tall architectural structure including an observation deck for long-distance sightseeing. The height of the deck corresponds to the distance and elevation of the view. The logical relation imported for the blend is thus DECK (floor, altitude) that binds together the number of floors one has climbed and the corresponding elevation of that floor. A digital display mounted on the wall inside the building, or just a mobile AR app, showing a print in the series would stimulate the person's imagination of a third story. The person climbs the stairs inside an observation tower. When reaching the observation deck on the target floor, one sees the view of Mount Fuji from a distance. Climbing to different floors is mapped with climbing in different towers or to different decks, acquiring different views. Table 5.13 summarizes the formal representation of the blending processes.

100 Famous Views of Edo (by Hiroshige)

Events: Each print in the series presents a scene in the city of Edo with the location written in the title. Each scene can be regarded as an event having taken place in Edo. The series forms a narrative of thematically related events.

Variation: Many elements of Edo repeatedly appear in the series, including Mount Fuji, bridges, rivers, boats, trees, women in kimono (a kind of traditional Japanese costume), and others. Each print is like a unique combination of the selected elements. Quite some prints feature a similar visual composition with a pronounced foreground object in front of the spectacular view.

Collection: The series is a collection of 119 prints together with an extra for table of contents, which are arranged in four sections according to season. Spring is the largest section made up of 42 prints. The number of prints in each section decreases from 30 in the summer, 26 in the autumn, to lastly 21 in the winter. Table 5.14 summarizes the characteristics of a few prints from each season.

Content: The prints present a wide array of scenes captured in Edo, ranging from those in the streets, temples or shrines, or gardens, to bird's-eye views over the city with Mount Fuji in distant background or during festival. With locations clearly written in the titles, the prints can also be grouped by region in today's Tokyo. They can be categorized into the city center (e.g., Nihonbashi), the north (e.g., Ueno, Asakusa, and Sumida River), the west (e.g., Shinjuku), and the south (e.g., Meguro and Shinagawa).

Framing: The series prints, in stark contrast to other preceding ukiyo-e prints, were designed in the portrait format. The visual composition of each print simulates a viewing angle taken at the named location at a certain height, ascending from the street level to the bird's-eye view level. Quite some pieces repeat a similar visual framing. For example, the street views of "Wholesalers of Cotton Fabrics at Ōtemma Street" (spring) and "A Drapery Store at Ōtemma Street" (autumn) are symmetrical. Both "Kameido Ume Garden" (spring) and "Iris Flowers at Horikiri" (summer) feature plants in the foreground of street-level views. Both "Suidō Bridge at Suruga Terrace" (summer) and "Tanabata Festival in a Prosperous City" (autumn) feature festival flags in the foreground, bird's-eye views of the city in the middle ground, and Mount Fuji in the background. In front of the bird's-eye views in both "Suijin Woods, Inlet, and Sekiya village viewed from Massaki" (spring) and "Asakusa Paddy Fields and Pilgrims to Ōtori Shrine" (winter), stands the foreground of an indoor space.

Synopsis: The series visually narrates happenings in the city of Edo, which is the story world. Streets, temples, shrines, gardens, bridges, seasons, and festivals are the settings. Visitors, pilgrims, and residents are characters, together with many objects and animals. An implicit and imaginary character visits different locations in Edo, identifying different objects and people in the setting, visually arranging them in a stylized pattern, and composing an event taking place at the location.

Formal representation for mapping: Table 5.15

Suggested routines: The series can be regarded as a location-based narrative in which events took place in different locations in Edo. Clearly written in the title of each print, the location of each event can be matched with their respective spots in today's Tokyo. It is found that all the events are evenly distributed across the central districts, the east, the north, the west, and the south. Meanwhile, each print picks up varied elements but arranges them in a pattern reminiscent of a view

TABLE 5.14 Selection of prints from each season of *100 Famous Views of Edo* with the corresponding content and viewing angle

Print title	Content	Viewing angle
Spring:		
Wholesalers of Cotton Fabrics at Ōtemma Street	Street, women in kimono, shop	Street-level view
Suruga Street	Street, shops, buyers, transporters, Mount Fuji	High-angle view
Kameido Ume Garden	Plum trees, visitors	Street-level view with part of a plum tree in the foreground
Suijin Woods, Inlet, and Sekiya village viewed from Massaki	Window frame, plum tree, river, boats, raft, mountains	Bird's-eye view with a circular window frame in the foreground
Summer:		
Suidō Bridge at Suruga Terrace	Koinobori (traditional windsock shaped like koi fish for festival), river, bridge, Mount Fuji	Bird's-eye view with a pronounced koinobori waved in the foreground
Pagoda of Zōjōji Temple and Akabane	Temple, river, bridge, street	Bird's-eye view with part of the temple in the foreground
Thunderstorm at Ōhashi Bridge and Atake	Rain, bridge, travelers, river, raft	Bird's-eye view
Iris Flowers at Horikiri	Flowers, stream, visitors	Street-level view with pronounced flowers in the foreground
Autumn:		
Tanabata Festival in a Prosperous City	Traditional festival flags, houses, bamboo straws, Mount Fuji	Bird's-eye view with a pronounced flags waved in the foreground
A Drapery Store at Ōtemma Street	Street, shop, parade	Street-level view
Winter:		
Kinryūzan Temple at Asakusa	Lantern, temple, pilgrims	Street-level view with a lantern and part of a gate in the foreground
Asakusa Paddy Fields and Pilgrims to Ōtori Shrine	White cat, window frame, paddy fields, pilgrims, white Mount Fuji	Bird's-eye view with a white cat beside a window frame in the foreground

through a window. The variation in the viewing location and the consistency in the framing render the series applicable to mapping with maintenance routines that entail changing places within a city, a town, or a district. Consider a working-class person whose job duty entails travels to different locations in a city, say, for meeting clients, delivering goods, or giving people rides. He or she might need to have a

TABLE 5.15 The formal representation of events in the narrative of *100 Famous Views of Edo* for mapping with routines

Iterative processes	*Successive outcomes*
...	
VISIT (Kameido)	
VIEWPOINT (street-level view, spring)	APPEAR (plum trees, visitors)
...	
VISIT (Suruga Terrace)	
VIEWPOINT (bird's-eye view, summer)	APPEAR (koinobori, river, bridge, Fuji)
...	
VISIT (Ōtemma)	
VIEWPOINT (street-level view, autumn)	APPEAR (street, shop, parade)
...	
VISIT (Asakusa)	
VIEWPOINT (bird's-eye view, winter)	APPEAR (white cat, window frame, paddy
...	fields, pilgrims, Fuji)

takeout lunch in different places. Sometimes, an office worker or a college student might also need to enjoy a takeout in a park, on the campus, or in other public spaces. Takeout food can be boring and repetitive, but it is convenient and sometimes necessary. Imagine someone buying a takeout meal and then walking north to find a bench in a park. While enjoying the meal on a spring day, they virtually see a street-level view like that depicted in "Kameido Ume Garden." When having a takeout on a rooftop during summer, the high-angle view of "Suidō Bridge at Suruga Terrace" appears. On an autumn day, the person may sit at a corner in the square and then virtually see a street view like "A Drapery Store at Ōtemma Street." When it turns chilly in winter, the person stays in the office to eat a packed lunch. When he or she looks toward the window, the interior plus outdoor view of "Asakusa Paddy Fields and Pilgrims to Ōtori Shrine" appears. To sum up, the location and altitude where one eats a takeout, together with the season or weather, determine the corresponding view in the series to be shown.

Imported relations for blending: In the series, the imaginary character visits different locations in Edo. They start with Nihonbashi, the city center, followed by going north, south, east, and west to many other locations. For those street-level views, the character captures the scene straight away after arriving at the location. To attain some bird's-eye views, they have to look for a higher place in the surroundings, which can be a terrace, a tower, or a hill. After getting to a perfect spot to appreciate the spectacular view, they might take out a packed meal or snack to eat. In real life, the working-class person brings a takeout lunch and looks for a comfortable "seat" in a public space to eat it. While eating, the person might also aimlessly look around for casual enjoyments. Both sightseeing and having a takeout involve finding and taking an available and favorable seat in a public space, where anyone can enjoy the surrounding views. The external frame [public space]

TABLE 5.16 The formal representation of the blends between having lunch on workdays and *100 Famous Views of Edo*

Having lunch on workdays	Selective projection and pattern completion	100 Famous Views of Edo
…	…	…
BUY (person, takeout) →	BUY (person, takeout)	VISIT (Kameido)
GO (person, north) →	GO (person, north)	
	VISIT (Kameido)	
	[public space]:	
	PARK (seat, street-level view)	
FIND (person, seat) →	FIND (person, seat)	
	VIEWPOINT (street-level view, spring)	← VIEWPOINT (street-level view, spring)
	APPEAR (plum trees, visitors) ←	APPEAR (plum trees,
EAT (person, takeout) →	EAT (person, takeout)	visitors)
…	…	…
BUY (person, takeout) →	BUY (person, takeout)	VISIT (Suruga Terrace)
GO (person, east) →	GO (person, east)	
	VISIT (Suruga Terrace)	
	[public space]:	
	ROOFTOP (seat, bird's-eye view)	
FIND (person, seat) →	FIND (person, seat)	
	VIEWPOINT (bird's-eye view, summer)	← VIEWPOINT (bird's-eye view, summer)
	APPEAR (koinobori, river, bridge, Fuji) ←	APPEAR (koinobori, river, bridge, Fuji)
EAT (person, takeout) →	EAT (person, takeout)	
…	…	…

includes a few spatial relations, such as PARK (seat, view), OLD TOWN PLAZA (seat, view), TERRACE (seat, view), or ROOFTOP (seat, view), that bind a position in space to a view. They can be imported to blend sightseeing and eating a takeout in a public space. When the person is having a takeout lunch in a park or on a rooftop, a mobile AR app displays a view of Edo according to one's current location and altitude in the real-world city. One might imagine a third story in which he or she is traveling to an old town in Japan and enjoying the view with the packed meal. Table 5.16 summarizes the formal representation of the blending processes.

53 Stages of the Tōkaidō (by Hiroshige)

Events: Tōkaidō, literally meaning "east sea road," is the name of a linear road connecting Edo (the new capital) and Kyoto (the old capital) along the eastern coast of Honshū, the main island of Japan. It starts from the center of Edo, Nihonbashi, through 53 stations (in Japanese, Shukuba, which is a place for travelers to rest or stay), to Sanjō Ōhashi in Kyoto. In this work, each print orderly presents a road scene and is named after a station explicitly. Each scene can be seen as an event taking place at a station on the road. The series thus forms a narrative of spatially ordered and thematically related events.

Variation: The road was one of the busiest in the Edo period. Most scenes depicted in the series feature travelers on the road. Most of them travel on foot, while very few ride a horse or travel in a litter with transporters. The settings along the road include the seashore, boats, rivers or streams, bridges, villages, shops, fields, and hills. The prints seem to repeatedly re-arrange these elements and present them in a myriad of ways.

Collection: According to the order of the stations, this series, unlike *36 Views of Mount Fuji* (by Hokusai) and *100 Famous Views of Edo* (by Hiroshige), consists of 55 prints in a very specific order, with a starting point, a destination and 53 stations in between.

Content:

As the road runs along the coast, most scenes cover the seashore with or without boats (Station No. 1, 3, 8, 10, 16, 18, 30, 31, 32, 39, 42, and 43), other times a river or stream with or without bridges (Station No. 2, 4, 5, 6, 10, 12, 17, 19, 21, 23, 24, 26, 28, 34, 38, and 49). At times, the road runs across a village (e.g., Station No. 11, 15), past shops or restaurants (e.g., Station No. 20, a famous restaurant still in operation today), paddy fields (Station No. 27, 29), or hills (Station No. 16, 25, 46) too. Rain (Station No. 45 is renowned for its aesthetic and technical achievements) and snow (Station No. 15, 46) occasionally pop up in several prints as weather sometimes can turn nasty on the road.

Framing:

Most road views are presented from a street-level perspective, or from a slightly higher angle taken on a slope along the road. In a few exceptions, the prints present bird's-eye views (Station No. 18, 23, 24, 34) from a relatively high hill top.

Synopsis: The series successively and visually narrates happenings along the road from Edo to Kyoto. The characters are the varied travelers on the road. The events seem to include walking on the road or a bridge, riding on a horse, crossing a river on foot, sitting in a litter or on a boat, transporting parcels, purchasing goods, or resting. An implicit and imaginary narrator (or character) travels on the road, taking note of other travelers and their activities at each station, visually composing a road scene presented on each print.

Formal representation for mapping: Table 5.17

Suggested routines: The series can be seen as a kind of "road movie" in which, instead of traveling by car, the characters traveled mainly on foot in the Edo period. The story can be straightforwardly mapped with the daily walking routine. The stations on the road were thoughtfully planned and developed for the old-time travelers to rest or stay at night. Thus, successive stations are within daily walking

TABLE 5.17 The formal representation of events in the narrative of *53 Stages of the Tōkaidō* for mapping with routines

Iterative processes	Successive outcomes
…	
ARRIVE (Fujisawa)	
VIEWPOINT (street-level view, No. 6)	APPEAR (stream, bridge, travelers, torii)
…	
ARRIVE (Kanbara)	
VIEWPOINT (street-level view, No. 15)	APPEAR (snow, travelers, huts)
…	
ARRIVE (Yui)	
VIEWPOINT (high-angle view, No. 16)	APPEAR (cliff, seashore, boats)
…	
ARRIVE (Mariko)	
VIEWPOINT (street-level view, No. 20)	APPEAR (restaurant, travelers)
…	
ARRIVE (Shōno)	
VIEWPOINT (street-level view, No. 45)	APPEAR (rains, travelers, shrubs)
…	

distances, ranging from five to fifteen kilometers from each other, and the distance can be finished on foot in approximately two to six hours. Travelers or transporters of the old days could leave a station in the morning and arrive at the next station before sunset. Consider someone in the modern day who uses a pedometer or the mobile phone to count their steps or record the distance they walk. When they have finished the first 7.8 kilometers (or the first two hours of walking), he or she virtually sees the street scene at the first station, Shinagawa. When the person continues to walk for another 9.8 kilometers, one virtually arrives at the second station, Kawasaki. In fact, a Japanese mobile step counter app, 東海道五十三次歩計,[12] has been developed based on that concept. It shows which station the user has virtually arrived at according to the steps counted, yet it has no connection with the ukiyo-e prints. It would be better if the user would be rewarded with a virtual view based on the corresponding print after achieving a prescribed number of steps. The Fitbit fitness app, coming with the smart wearable device, rewards its users with virtual badges that seemingly respond to this call. According to a user's increasing distance walked since using the device, the badges given present landmarks or scenery spots from different places, including London, Hawaiian Islands, Serengeti, Italy, New Zealand, and many others.[13] However, spatial arrangements of these places are more comparable to that of the famous views of Edo than the stations on the Tōkaidō road. They are not spatially ordered on a linear travel route, which makes the mapping with walking less perfectly aligned. Conversely, the linearly ordered stations on the Tōkaidō road match the daily walking routine well. While the walking routine is a straightforward match, another maintenance routine, important but often overlooked, that should be considered is doing the laundry. Imagine someone doing the laundry weekly at home, including the tedious processes of bringing dirty laundry to the washing machine, setting the program, and folding clean and dry laundry. In the first week, one virtually stays at the first station, Shinagawa. With the laundry done, one becomes ready to move on the journey to the second station, Kawasaki, and virtually sees the scene from the corresponding print. Every laundry done means that the person can pack the travel clothes in the luggage and continue the journey to the next station on the road. After 53 weeks of laundry done, one virtually sees the destination, Sanjō Ōhashi in Kyoto.

Imported relations for blending: In the series, the imaginary character travels on the road in the Edo period. As they travel on foot from station to station, they can only pack a few spare clothes in the luggage and have to do the laundry by hand every night. When the laundry is dry the next morning, they then pack it in the luggage again and move on. In reality, a person might wait and do laundry once a week. They take the dirty laundry to the washing machine. After the clean laundry is dry, they fold the clothes and prepare to wear them in the coming week. To connect doing laundry on the road and doing laundry at home, the external frame [accommodation] can be invoked. It refers to places for staying temporarily. Many guesthouses of today provide washing machines. If not, one can still do laundry by hand during travel. The new relations that can be imported from [accommodation] are CHECKIN (guesthouse) and WASH (clothes, guesthouse), which mean a person checking in a guesthouse can use the machine to wash their own clothes. They bind together doing laundry in a guesthouse and doing laundry at home. With a digital display near the washing machine at home showing the road scene at a station on the Tōkaidō road, the person doing laundry at home can imagine a third story about doing laundry when staying in a guesthouse on the road. Once the laundry is done, they can wear clean clothes in the coming week and continue the journey to the next station. Table 5.18 summarizes the formal representation of the blending processes.

Summing Up

Behind different routines lie different reasons, which require different kinds of stories to match. As routines are actions or behaviors regularly performed, matching stories should basically feature a sequence of events that repeat some components like actions or happenings. For goal-directed routines, people are concerned with progress, and successive events should show progression through growth in size or piling up of items. Exemplary matching stories can be found in huge assets of chain tales, which cover three types of goals in different routines. First, tales with plots in the "the quest" or "overcoming the monster" type tend to match routines performed in pursuit of a definite goal, like daily workouts to build muscles or taking medications to strengthen the immune system. Second, tales featuring the "voyage and return" plot are promising for mapping with routines aiming at a checklist of

TABLE 5.18 The formal representation of the blends between doing laundry at home and the narrative of *53 Stages of the Tōkaidō*

Doing laundry	Selective projection and pattern completion	Narrative of 53 Stages of Tōkaidō
...	...	...
	ARRIVE (Fujisawa)	← ARRIVE (Fujisawa)
	VIEWPOINT (street-level view, No. 6)	← VIEWPOINT (street-level view, No. 6)
	APPEAR (stream, bridge, travelers, torii)	← APPEAR (stream, bridge, travelers, torii)
	[accommodation]	
	CHECKIN (guesthouse)	
	SAME MACHINE (home, guesthouse)	
WASH (clothes, home) →	WASH (clothes, guesthouse)	
FOLD (clothes, home) →	FOLD (clothes, guesthouse)	
PACK (clothes, drawer) →	PACK (clothes, luggage)	
	CHECKOUT (guesthouse)	
...	...	...
	ARRIVE (Kanbara)	← ARRIVE (Kanbara)
	VIEWPOINT (street-level view, No. 15)	← VIEWPOINT (street-level view, No. 15)
	APPEAR (snow, travelers, huts)	← APPEAR (snow, travelers, huts)
	[accommodation]	
	CHECKIN (guesthouse)	
	SAME MACHINE (home, guesthouse)	
WASH (clothes, home) →	WASH (clothes, guesthouse)	
FOLD (clothes, home) →	FOLD (clothes, guesthouse)	
PACK (clothes, drawer) →	PACK (clothes, luggage)	
	CHECKOUT (guesthouse)	
...	...	...

items or tasks. Third, tales of the plot type "rags to riches" match routines oriented toward long-term or distant goals, like trading used items or recycling used materials for economic or environmental sustainability. Different types of goals entail different forms of progress indicators in the matching stories. Progress made toward a definite goal can be represented by a rise in quantity (e.g., higher percentage) or quality (e.g., better substance). It reversely informs distance to the endpoint on a linear scale. Long-term goals are usually hard to be tied with a definite endpoint, and so the distance remaining cannot be known. Instead, a sense of making progress is delivered through constant growth of one attribute (e.g., power). For checklist goals, the number of items or tasks in the list still left undone can be used to indicate the progress made. In short, the progress indicators for the three types of goals, respectively, emphasize distance to the endpoint (definite goals), outstanding items (checklist goals), and an ever-growing attribute (long-term goals).

In maintenance routines, people prefer seeing variety rather than progress, which requires another category of stories for mapping. This kind of story features a series of thematically related events that not only repeat certain patterns but also feature constant variations. Many ukiyo-e print series demonstrate consistency in theme, style, and visual composition, while being varied in arrangements and visual details. Most importantly, some of them can be interpreted as a simple narrative of events with each print showing one event. Due to different themes and contents, different print series can be considered for mapping with different maintenance routines, such as stair climbing, having a takeout lunch, or doing the laundry. It should be noted that goal-directed routines and maintenance routines are not always separate. Some routines driven by long-term goals in particular can be seen as maintenance routines, for example, trading used goods to support a lifestyle with economic sustainability. The crux here is not to draw a solid line between different kinds of routines, but to prioritize the distinction between progress and variety. If the person practicing a routine is keener on progress, chain tales should be considered for mapping; if one prefers variety, theme-based narratives would be more appropriate. Figure 5.2 summarizes the distribution of the case studies in this chapter over the continuum between variety-focused and progress-focused iterative processes.

Whether an analogous story is a chain tale or ukiyo-e series narrative, mapping and blending with a routine requires formal representation of the event sequence, invocation of an external knowledge frame, and importing new relations into the blend. Through case studies of a list of exemplary chain tales and ukiyo-e series, a wide range of routines, from goal-directed routines associated with progress to maintenance routines benefited from variety, can be blended with an analogous story to yield a third imaginative story that motivates. Yet, most of these analogous

maintenance routines		goal-directed routines			
variety-focused process		long-term goals	checklist goals	definite goals	progress-focused process
Theme-based narratives		The Stonecutter (rags to riches)		The Turnip (the quest)	Chain tales (plot-type)
	36 Views of Mount Fuji		The Flying Ship (voyage & return)		
	100 Famous Views of Edo			The Three Little Pigs (overcoming the monster)	
	53 Stages of the Tōkaidō		Tsarevitch Ivan, the Firebird and the Gray Wolf (voyage & return)		

FIGURE 5.2 Distribution of analogous stories over the continuum between variety-focused and progress-focused routines.

stories are relatively short compared with the lifespan of many corresponding routines. In "The Turnip," overall six family members were summoned to pursue the goal, so that it can be mapped with six workout sessions, while a muscle-building workout plan is definitely longer. In "The Three Little Pigs," the pigs consecutively built three houses before taking down the villain, meaning it may be mapped with three days of medication, but prescriptions can last much longer today, not to mention the never-ending task of taking health supplements. Furthermore, a person may set another goal after achieving one goal, for example, enrolling a new weekly course after finishing the previous one. In case of long-term goals, the analogous stories are required to be extendable or flexible in length. Although each ukiyo-e series has dozens or even over 100 prints, maintenance routines can last indefinitely. All in all, to adapt the diversity and dynamic of daily living, analogous stories for routines should be further elaborated. Expressive iteration not only applies to the event level but also to the story level. In the next chapter, I will discuss the next level of iteration.

6

ITERATION OF STORIES

Aligning and blending a routine with an analogous story can motivate a person in two ways. On the one hand, gradual increase in quantity or quality, or piling up of items along successive events in the analogous story represents tangible progress, which satisfies the basic psychological need for competence, particularly in case of a goal-directed routine. When a person practices a routine in pursuit of a definite goal like running every morning to lower blood cholesterol level, a checklist goal like attending every class in a short course, or a long-term goal like trading used goods for economic sustainability, they are interested in seeing themselves getting closer to their respective goal, that is, bringing their blood cholesterol level to a normal range, getting a checklist of tasks done, or achieving personal growth step by step. Different categories of chain tales offer analogous stories together with progress indicators appropriate for different goal-directed routines. On the other hand, variations between successive events in analogous stories can fulfill another psychological need for novelty, which is more relevant to a person engaging in what I call a maintenance routine. When one regularly climbs the stairs instead of taking the elevator, eats a takeout lunch on workdays, or does the laundry at home, they would prefer some kind of variations introduced to the routines because that leads to novel experiences and makes the routines less boring. Some ukiyo-e print series can be interpreted as theme-based narratives with consistency and variety that match those maintenance routines. However, routines in real life typically last for a substantial period of time, which can be longer than the chains of repetitive events in most chain tales and the number of prints in ukiyo-e series. For example, a daily running routine to lower blood cholesterol levels typically takes 24 weeks for observable results,[1] while a matching story for it, such as "The Turnip," consists of six repetitive events, far from enough to cover the whole course (even when each newly joining family member is mapped to one week of the routine). A short

DOI: 10.4324/9781003391449-8

introductory course of a foreign language might consist of 12 weekly sessions. Mapping each session with one team member in the analogous story "The Flying Ship" is sensible, but there are only eight members on board in total. One simple solution to this mismatch problem is to lengthen the story by adding events to the chain. For instance, one could imagine asking more family members like other siblings or pets to pull the turnip out of the soil; more strangers with other supernatural powers, say being able to become transparent or restore broken things, might show up and board the flying ship. This is of course an ad-hoc solution. Adding similar events indefinitely to a chain is inconvenient and awkward.

A routine is the iteration of an action in daily life. A story matching a routine should consist of iteration of a comparable event in a fictional world. Real-life action can be mapped with a fictional event. If the iteration in reality lasts longer, that is, involving more cycles than the iteration in the story, one might consider repeating the story with some varied content, in other words, iteration of the whole story. For example, the original story of "The Turnip," which iterates the act of pulling out the turnip, can be followed by a sequel with a similar "quest" but on a different goal, such as pulling out another gigantic root vegetable, such as a carrot, beet, or celeriac. A sequel is similar to the iteration of a story with repeated story structure and characters but a new goal or villain. One or more sequels can form a series, or a collection of "The Turnip" and similar tales, which can be mapped with goal-directed routines that continue for weeks or months. The development from a sequel to a series is commonplace in films, such as *The Matrix* (1999–2021), *Infernal Affairs* (2002–2003), and *Mission: Impossible* (1996–present). In the film industry, sequels are only possible when the predecessor has enjoyed substantial box office success. Examples of sequels leading to series also exist in other forms of creation, including novels, television shows, and comics. The strategy can be applied to series of analogous stories for mapping with long-running routines. In fact, sequels to well-received ukiyo-e print series were published in the Edo period too, resulting in series of travel narratives (as each print series can be regarded as a travel narrative). In this chapter, I will draw references from sequels in popular culture and fine art to illustrate how a series of analogous stories can be developed by creating sequels of an archetypal story.

A series developed from the sequels of a chain tale can be an episodic series like *Bones* (American TV series, 2005–2017) or *Galileo* (Japanese TV series, 2007–2013) in which each episode tells a self-contained story about the same set of main characters in the same fictional world. For instance, the peasant family in "The Turnip" continues in the sequels to persistently harvest their gigantic root vegetables. On the other hand, a series developed from sequels following a theme-based narrative can be an anthology series like *Black Mirror* (British TV series, 2011–present) that presents different stories with different characters in different settings. For instance, different ukiyo-e print series show different travel narratives taking place on different roads between Edo and Kyoto. Both episodic and anthology series extend the themes and tones of the original stories with a balance of consistency and variety but showing limited or no progression. One shortcoming of mapping a series of analogous stories with a long-running routine is the lack of continuity between successive stories in the series, which may lead to casual skipping or quitting. A person who has finished one self-contained story might not feel the urge to move on to the next and so easily procrastinate. To enhance continuity in a series, two literary techniques commonly used in varied narrative forms can be considered. The first is a frame story, which refers to a story that contains other stories within it. Some notable examples include *The Decameron, Canterbury Tales,* and *1001 Nights*. In each of these novels, a series of short stories, usually organized in different sub-themes, were narrated by one or more characters in another fictional world. After finishing one short story, a reader is transported back to the frame story, which then leads the reader to another short story. The reader who follows the frame story would also like to know what the next short story is about. Adding a frame story to a collection of analogous stories would give a person more reasons for continuing with the routine, not just because of the novel experience that emerged from the variety of each analogous story but also due to the unfinished frame story. The second technique to boost continuity is weaving an overarching storyline through all analogous stories in the series. Each analogous story introduces and concludes its self-contained storyline, meanwhile advancing the large story arc incrementally. A person practicing the routine would find each analogous story novel and satisfying, as it has its own beginning and an end. It also seems to contribute to a part of the continuing main narrative, where loose ends and mysteries are yet to be resolved. This is a common strategy used in novels, films, television shows, or comic series that assume the characteristics of serial narratives. Illustrative examples include the film serial *The Perils of Pauline* (1914), the Japanese tokusatsu (literally meaning "live action with special effects") television series *Kamen Rider* (1971), and many Japanese manga series like *JoJo's Bizarre Adventure* and *Saint Seiya*.

In Chapter 5, I introduce two kinds of analogous stories for mapping with different routines, namely, chain tales for goal-directed routines and ukiyo-e travel narratives for maintenance routines. The former features progression, while the latter foregrounds variation. Starting with an analogous story that matches a target routine, we can develop a sequel by iterating some characters or settings of it and put together a collection of chain tales like an episodic series, or a collection of travel narratives like an anthology series. There are two different techniques to strengthen the continuity in both types of series. To illustrate the first technique, adding a frame story, in addition to the aforementioned classical references, a type of traditional Chinese winter calendar charts will be presented and connected to the "frame" metaphor in storytelling. Regarding the second technique, weaving an overarching storyline, apart from many examples from novels, films, and comics, a peculiar and notable work of serial art will be discussed in relation to several characteristics of serial narratives, including iteration of recognizable features, multiple storylines, and momentum in the main narrative.

From Sequels to Episodic Series

A "sequel" typically continues or expands upon the story of an earlier work. The events in the sequel generally follow the original story chronologically, and it becomes a "prequel" if the events precede the original. As a sequel is usually made in response to a successful work, it tends to imitate the original but cannot simply copy it. It attempts to keep the essence of the original, meanwhile imbuing it with new meanings. Richard Pfefferman (2013) appreciates the creativity in these "replications" (including spin-offs, adaptations, and remakes) and distinguishes them as "reinventions" (1–3). In this sense, sequels and similar reinventions embody the core of expressive iteration on the story level, that is, iterate stories for new meanings. By examining the synopses of a few successful (both in terms of box office and fans' responses) Hollywood sequel movies, one can uncover the iteration therein. In *Back to the Future* (1985), the teenage protagonist accidentally travels to 30 years ago in a DeLorean-turned time-traveling vehicle built by his eccentric scientist friend. In 1955, the teenager inadvertently intervenes in his parents' romantic relationship and consequentially affects his own existence. He has to fix it and get back to 1985. In the first sequel, *Back to the Future Part II* (1989), the teenager follows the scientist on an upgraded DeLorean mobile to 30 years later to save his future son. As the story unfolds, the teenager unexpectedly allows his father's enemy to change the history. The teenager has to go back to 1955 again and revert everything. It iterates many events of his parents' romance story in which the protagonists even saw the versions of themselves in the first movie, visually representing recursion. The sequel repeats the "voyage and return" plot in the original, retains the accomplishments the protagonists had in 1955 as shown in the first movie, and adds new challenges for them to tackle in the updated 1955. Similar iteration but of a different plot type can be found in *The Terminator* (1984) and *Terminator 2: Judgment Day* (1991). In the original, a human soldier and a humanoid killer are separately sent from 2029 to 1984. The humanoid is programmed to kill a woman, who would give birth to a future human leader against the machine. Meanwhile, the soldier's mission is to protect this woman to save the world. The sequel repeats the "overcoming the monster" plot of having one protector and one killer sent from future but updates the protector to be the humanoid that appeared last time and the killer to be a more advanced killing machine.

Other film series that features iteration of "overcoming the monster" include *Alien* (1979) and *Aliens* (1986). The sequel repeats and escalates the brutal fights between humans and the deadly monsters, with the humans more armed and the monsters more numerous. The *Indiana Jones* film series is a prominent example of iterating "the quest" plot. From *Raiders of the Lost Ark* (1981), *Indiana Jones and the Temple of Doom* (1984), to *Indiana Jones and the Last Crusade* (1989), Jones repeatedly sets off on a journey with companions to search for a different but similarly mythical, miraculous treasure. His companions are different in every journey but gradually grow in size from a female partner in the first film, plus a 12-year-old boy in the second film, to a friend, a new acquaintance, and most importantly his father in the third installment of the series. Iteration can also be observed in film series without time travel and magical powers but with only human will power and determination. The *Rocky* film series features iteration of the "rags to riches" plot. The first film *Rocky* (1976) starts with the protagonist, Rocky, being a small-time boxer, who happens to get a chance to fight the world heavyweight champion, Apollo. Rocky loses the match but wins the respect from the masses. *Rocky II* (1979) follows the same protagonist who struggles to earn a living and has to resume boxing. Rocky wins in a rematch with Apollo. In *Rocky III* (1982), Rocky enjoys wealth and fame and wants to retire. With a new contender, Rocky loses his trainer friend and the fight. His former rival Apollo trains him and he wins again. In *Rocky IV* (1985), Apollo is killed in a boxing match with a Russian boxer and Rocky wants revenge. With the help of many people and his determination, Rocky defeats the Russian boxer for his friend. Across the series, Rocky goes through a series of wins and losses, and then wins again, while facing ever-escalating challenges. To sum up, iteration of stories of varied plot types can be found in a list of well-known film series (see Table 6.1). Their treatments in repetition and variation across the sequels inform the ways that can be used to develop an analogous story, that is, a chain tale of a certain category as presented in Chapter 5, into an episodic series.

An episodic series consists of self-contained episodes. Each episode engages the same set of main characters, and its storylines do not continue across other episodes.[2] Strictly speaking, the characters do not usually exhibit personal growth or development. In this sense, only classical sitcoms like *Mr. Bean* or long-running comic strips such as *Peanuts*, *Garfield*, and Japanese manga *Doraemon* can be pure episodic series. In fact, many sitcoms in recent decades like *Friends* (1994–2004) and *The Big Bang Theory* (2007–2019) have accommodated subtle and gradual character developments (e.g., marriage, pregnancy, job change) across seasons. Similar developments can also be seen in long-running drama television series (and

TABLE 6.1 A summary of iteration of stories in selected film series spanning different plot types

Series	Plot type	Repeated elements	Varied elements
Back to the Future	Voyage and return	Strange time in the past, time-travel vehicle	More complex challenge
Terminator	Overcoming the monster	Threat, fight	Stronger protagonist, stronger monster
Alien	Overcoming the monster	Threat, fight	Stronger protagonist, more monsters
Indiana Jones	The quest	Adventure	New goal, new or more companions
Rocky	Rags to riches	Gain, loss	New contender

occasionally film series such as *Rocky*). In *Bones* (2005–2017), the protagonist, a forensic anthropologist with an exceptionally high intelligence quotient score, develops a romantic relationship with her long-time working partner in Season Six, followed by having a daughter together, getting married, and giving birth to a son in the succeeding seasons. In the Japanese drama *Doctor-X: Surgeon Michiko Daimon* (2012–2019), the talented, self-assertive yet eccentric freelance surgeon works in different hospitals in different seasons and deals with different management teams. In short, episodic series in today's popular culture, particularly those with multi-season runs, might have the settings evolved or changed along with time. This gives a sense of resonance that the characters grow and advance in sync with the audiences' real lives. This fiction-reality synchronization applies perfectly to the mapping of analogous stories with routines as well.

When considering the way of developing an analogous story into an episodic series for a long-running routine, both film and television series may cast light on the matter. The list of film series in Table 6.1 suggests different ways of iterating a story of a certain plot type, which also refers to a category of chain tales for routine mapping. Some elements in the plot are repeated, while others can be renewed or escalated in intensity and scale. Long-running television series suggest changes in settings or character developments along with time. In the following sections, the exemplary chain tales for mapping with different goal-directed routines as discussed in Chapter 5 are elaborated into corresponding episodic series as demonstration.

The Turnip and Other Root Vegetables

"The Turnip" belongs to the category of chain tales that exhibit "the quest" plot type and matches goal-directed routines targeting definite goals, such as exercising to achieve an optimal body weight. It can be elaborated into a tale series with reference to the repetitions and variations found in the *Indiana Jones* series. In the film series, the same protagonist, Indiana, sets out on a similar adventure journey in each movie in pursuit of a precious item, with varied or additional companions. The protagonist in "The Turnip," the grandfather from the peasant family, aims to harvest the gigantic turnip, with the help of increasingly more family members. In a possible sequel, the same grandfather might find another gigantic root vegetable, say a carrot, deep in the soil of his farm. He cannot pull it out alone, and so he repeatedly asks family members to join the quest. New members, in addition to the original family, might include a nephew, the granddaughter's new classmate, a new neighbor, or other animals like a fox or squirrel. For a tale series, many gigantic root vegetables can be considered, not only a carrot, but also a beet, celeriac, and potato.

As informed by long-running television series, changes in the peasant family setting along the passing of time can bring new elements to the tale series. For instance, different root vegetables can be harvested in different seasons. A quick

TABLE 6.2 A possible series of "The Turnip and other Root Vegetables"

Season	A root vegetable goal	New members	Updated setting
Winter	A gigantic turnip	The grandfather, the grandmother, the granddaughter, the dog, the cat, and the mouse	The farm
Spring	A gigantic celeriac	A fox or a squirrel	The mouse having meet new animal friends
Summer	A gigantic carrot	A nephew	Relatives coming to visit
Fall	A gigantic beet	A new neighbor	A new neighbor moving to the area

search in the Internet results in related information. Turnips and celeriac are harvested in the fall and winter season. Carrots are harvested in June and August, beets in September and October, and potatoes from August to October. The peasant family can plant and harvest different root vegetables all year round. They can also have a nephew visiting in summer, the granddaughter might meet a new classmate in new semester, or a new neighbor might move to the area. Table 6.2 summarizes the ideas and delineates a possible series for "The Turnip and other Root Vegetables."

The Three Little Pigs and Other Threats

"The Three Little Pigs" also belongs to the category of chain tales that are good for mapping with goal-directed routines. Because of its plot type "overcoming the monster," it matches better those with an avoidance goal, such as taking pills to cure diseases and strengthen immunity. The tale can be elaborated into a series as informed by the two aforementioned thriller film series, *Terminator* and *Alien*. In the *Terminator* series, the female protagonist survives the first humanoid's assassination attempt. In the sequels, not only is her physical strength boosted, but her protector and her killer also become stronger. To develop the chain tale into a series, one can imagine a sequel where the three pigs keep patching the loopholes in their home security, such as building fences, as they keep coming across more threatening predator, say, a brown bear. The bear is strong enough to repeatedly break different fences until the pigs build them with metal. In the *Alien* series, after the first deadly encounter with the alien, the protagonist becomes fully armed in the second round, but there are many more aliens this time around. It follows that the three pigs might need to prepare for the return of more wolves, which might come from the same pack of the one scalded in water in the original story. These wolves are able to climb over the metal fences. The pigs then have to build higher and higher fences.

TABLE 6.3 A possible series of "The Three Little Pigs and other Threats"

Season	A new threat	Upgraded defense	Updated setting
Spring	A wolf	A house built of straw, wood, or brick	The wolf blows the house down
Spring/Summer	A bear	A fence made of bamboo, wood, or metal	The bear, just after hibernation, breaks the fence
Spring/Summer	Three wolves	A fence built up to one, two, or three meters in height	The three wolves, from the same pack, climb over the fence
Fall/Winter	A polar bear	A house with a door made of different materials	The polar bear, hunting for food before winter, break into the house

The new threats might emerge in different seasons. One can assume that the original story takes place in spring, which is a good season for construction work. The first and second sequels might happen in spring or summer, when animals are actively searching for food. In fall or winter, the threatening animal coming to the pigs' house could be a polar bear so strong that even the metal fences cannot keep it out. The polar bear breaks the wooden door and enters their house. The pigs escape to their cousin's house with a metal door and survive the ordeal. Table 6.3 summarizes the ideas and delineates a possible series for "The Three Little Pigs and other Threats."

The Flying Ship and Other Companions

"The Flying Ship" belongs to the category of chain tales that demonstrates "voyage and return" plot type, which can be mapped with routines that entail the completion of a list of tasks. Examples include taking a three-month short course. The protagonist in this tale boards the flying ship, builds his team, accomplishes the impossible tasks given by the cunning king, and finally returns home with the princess. The flying ship is comparable to the time-traveling vehicle in the *Back to the Future* series. In the film series, the young protagonist also rides the vehicle (to another time not place), comes across unexpected new threats, fixes the problems, and safely comes back. In the film sequel, he rides the time-traveling vehicle again to fix another problem in his future family, but his action triggers some chain reactions that lead him back again to 1955 when his parents met. Consider a similar setup for a sequel to the chain tale. The simpleton protagonist, after the first voyage and return, continues to travel on the flying ship with the princess. During their travel, a sorcerer sees the flying ship and wants to take possession of it. He lies to the princess that her young sister is seriously sick and that she should go home to visit her. The couple fly to the cunning king's castle again, only to find that

TABLE 6.4 A possible series of "The Flying Ship and other Companions"

Season	Similar strange setting	Unplanned team	Updated setting
One	The flying ship, the cunning king	The listener, the runner, the archer, the gobbler, the guzzler, the snowmaker, and the wood-carrier board the flying ship	The cunning king announces that his daughter will be married to the person who came with a flying ship
Two	The upgraded flying ship, the cunning king	The elastic man, the amazing chef, the invisible woman, the animal talker, and the restorer board the flying ship	A jealous sorcerer helps the cunning king and deceives the protagonist to come back to the king's castle

the sorcerer is now the right-hand man of the king to help him with his revenge. Fortunately, the protagonist has come with new companions on board this time. Inspired by Japanese manga or superheroes movies, the new team might include an elastic man who can stretch his body indefinitely, a chef who can chop down a tree with just one kick, a woman who can turn herself invisible, a man who can talk to animals, and a teenager who can restore any broken things to the original state. They defeat the sorcerer and meet the princess' sister. Table 6.4 summarizes the ideas and delineates a possible series for "The Flying Ship and other Companions."

The Stonecutter and More Wishes

"The Stonecutter" belongs to the category of chain tales that demonstrate "rags to riches" plot type. This category matches routines in pursuit of long-term goals, such as trading used items or money saving. The *Rocky* film series may shed light on how to develop the chain tale into a series. The film protagonist Rocky, an unknown boxer, manages to draw public attention after some respectful boxing fights. He, in the sequels, wants to stay low-key but then oftentimes resorts to taking on new challenges again. The stonecutter in the chain tale, after experiencing different transformations into different powerful beings, realizes the virtue of his original ordinary self. In a possible sequel, the stonecutter might come across another spirit, say from a river, who tempts him into another transformation journey. He wonders if he can make himself an extraordinary and satisfying life. He subsequently turns into a fish freely swimming in the river, a bird that skillfully catches a fish, thick fog that obscures the bird's eyesight, a gust of wind that blows away the fog, a windbreak wall around a castle that slows the gust, and once again a stonecutter who cut stones to repair the wall.

While the stonecutter meets the mountain spirit in the original tale on a hot summer day, he might meet the river spirit in a season when fog and strong winds are

TABLE 6.5 A possible series of "The Stonecutter and more Wishes"

Season	A new contender	New wishes	Updated setting
Summer	A mountain spirit	The stonecutter turning into a rich man, a prince, the sun, a cloud, a mountain, and then back into a stonecutter	The stonecutter worked on a hot summer day and felt tired
Late autumn	A river spirit	The stonecutter turning into a fish, a bird, fog, a gust of wind, a wall, and then back into a stonecutter again	The stonecutter worked in a stone factory and had to transfer water. He felt bored

both possible, say, the late autumn. In this sequel, the protagonist moves from the mountain to a stone factory to produce bricks from raw stones. For such purpose, he has to collect water from the river and that is how he meets the river spirit. Table 6.5 summarizes the ideas and delineates a possible series for "The Stonecutter and more Wishes."

From Sequels to Anthology Series

A chain tale analogous to a goal-directed routine can be iterated into sequels with both repeated and varied elements, forming like an episodic series. In an episodic series of chain tales, each tale is a self-contained story yet engaging the same set of main characters in a largely similar setting and featuring a very similar plot (or chain). In case of a theme-based story analogous to a maintenance routine, the characters and plot might not be explicit. Iteration of the story into a sequel leads to an anthology series. An anthology series typically puts together a collection of self-contained stories, which do not have any common characters or settings but do share a common theme. The British television series, *Black Mirror* (2011–2023), is a prominent example in recent years that explores different possible dystopian futures imbued with technology projected from the present day. Each episode in the series presents a different story with different characters in a different story world, but just under the same theme. Another popular genre of television series is documentary series, ranging from the Japanese travel series *The Silk Road* (1980–1981), the British nature series *The Blue Planet* (2001), and the long-running American reality show series *The Amazing Race* (since 2001), to the recent proliferation of miniseries on streaming video platforms exploring a variety of topics, such as *Street Food* (2019–present). Although a different terminology, a documentary series is also an anthology series of theme-based stories, with each episode focusing on separate sub-theme under a large common theme.[3] For instance, *Street Food* has run for three seasons since 2019 to explore street food around the world. The main theme of Season One is Asia, and the nine episodes walk audience through nine cities in the region, including

Bangkok, Oscar, Delhi, Yogyakarta, Chiayi, and others. Each episode is like a city tour, featuring a few street-food stores and the items they offer, as well as interviews with the store owners. In this sense, *Street Food* seems to echo the travel spirit of the ukiyo-e print series presented in Chapter 5.

Chapter 5 presents a few renowned ukiyo-e print series for mapping with maintenance routines. Each print series is in fact a theme-based narrative. Looking back on the renowned landscape prints in the final golden era of ukiyo-e, the main theme is largely travel, either on certain main road or in a region. Iterating a ukiyo-e print series into a sequel under the same theme leads to an anthology series of ukiyo-e theme-based narratives. The sequel does not need to repeat any characters, settings, or storylines of the original, but just maintains the same theme. In fact, the successful reception of many of these ukiyo-e print series in the Edo period has already engendered their sequels in those days. The following sections put together each famous print series with its sequel to showcase an anthology series of theme-based stories for mapping with often long-running maintenance routines.

53 Stages of the *Tōkaidō* (1833) and 69 Stages of the Kisokaidō (ca. 1837)

As mentioned in Chapter 5, *53 Stages of the Tōkaidō* is a series of 55 ukiyo-e woodblock prints published in 1833. Each print depicts a road scene around a station on the Tōkaidō, which is one of the main roads connecting Edo (the new capital) and Kyoto (the old capital). There are 53 stations on the Tōkaidō, together with Nihonbashi in Edo and Sanjō Ōhashi in Kyoto, resulting in 55 scenes of a travel narrative. The huge success of the print series led to a sequel initiated by the same publisher. In 1835, the publisher started to invite Keisai Eisen, another ukiyo-e artist, to work on a series on the theme of the Nakasendō (aka the Kisokaidō, a main part of it), another main road between Edo and Kyoto. Eisen stopped working on the series in 1837, and Hiroshige continued to finish it. At the beginning of the series, Eisen tends to depict dramatic moments among characters in the road scenes, including small talk between two geishas in Nihonbashi, a young travel assistant tidying up a horse's feces in Urawa, and a shop owner greeting an important guest in a litter in Kumagaya. In later prints, Hiroshige gradually shifted the tone to the enduring solitude on a long journey, oftentimes showing a line of silent travelers, such as in the prints titled "Yawata," "Ashida," "Wada," the famous "Tsumagome," "Nakatsugawa," and "Oi."[4] Eisen and Hiroshige are like two documentary film directors with nuanced styles directing different episodes of the same television series, which is not uncommon in today's production. The sequel series consists of totally 71 prints, including each of the 69 stations along the Nakasendō, the departure point Nihonbashi in Edo, and a double print of one particular station Nakatsugawa-juku. The Nakasendō is 533.9 kilometers in length, about 40 kilometers longer than the Tōkaidō, and has more stations. Unlike the Tōkaidō running along the eastern coast, the Nakasendō was built across the inland area. The sequel series thus shows more ups and downs in the mountains. At the beginning of the series, almost every print sees travelers on a relatively flat road. Starting from the sixteenth station, the roads become steeper, leading to something similar to hiking trails of today, such as "Tsumagome" and "Magome." Like the Tōkaidō series, the sequel series occasionally depicts scenes in the winter season. All in all, the sequel series iterates the travel theme, overlapping with the original series in certain road scenes like people crossing a river by boat (e.g., "Kawasaki" on the Tōkaidō and "Warabi" on the Nakasendō), passing by a paddy field (e.g., "Hara" on the Tōkaidō and "Mitono" on the Nakasendō), caught in a rainstorm (e.g., "Shōno" on the Tōkaidō and "Suhara" on the Nakasendō[5]), and braving in the snow at night (e.g., "Kanbara" on the Tōkaidō and "Oi" on the Nakasendō). Meanwhile, the two series differ in a lot of details on the roads including the characters, villages, trees, and landscapes.

In Chapter 5, I suggest that the Tōkaidō series matches the maintenance routine of doing the laundry well. When doing the laundry at home, a person might imagine doing it in a guesthouse while traveling on the Tōkaidō. Finishing laundry enables one to move on to the next station and finally reaching the destination, Kyoto. Yet, dirty laundry gradually accumulates again. One might iterate the travel story on another road, for example, the Nakasendō. Starting from Nihonbashi in Edo again, the person would go to the inland area this time and see more mountains than the seashore. Anyhow, the analogous story would bring them to another station on the Nakasendō every time they finish the laundry.

Famous Views of the Sixty-odd Provinces *(1853–1856) and* 100 Famous Views of Edo *(1856–1858)*

100 Famous Views of Edo, as mentioned in Chapter 5, is arguably Hiroshige's most highly regarded masterpiece in his final years. The series explicitly presents the famous attractions in the Edo city. It can be seen as a sequel to an almost immediately preceding work, *Famous Views of the Sixty-odd Provinces*. The two series share many commonalities. First, they both delineate a set of famous attractions in a region, with the preceding work spanning across provinces in Japan, while the succeeding one focusing on Edo. Second, both series consist of prints all in portrait format, which was atypical of other contemporary landscape ukiyo-e prints like Hokusai's *36 Views of Mount Fuji* or Hiroshige's own *53 Stages of the Tōkaidō*. It seems that Hiroshige aims to challenge the common practice of using the landscape format to present landscape. Third, using the vertical portrait format initially seems to be odd for landscape but in fact enables distinctively new framing possibilities in both series. Within a vertical frame, the artist was inclined to arrange objects closer to the viewer in the bottom and objects further from the viewer in the top part, like a bird's-eye view. Similar bird's-eye views can be found in many prints in both *Famous Views of the Sixty-odd Provinces* and *100 Famous Views of Edo*, such as "Suruga Province," "Izu Province," and many others in the former, as well as "Nihonbashi: Clearing after Snow," "Suruga-chō," and "New Fuji in Meguro" in the latter. Meanwhile, the groundbreaking style featuring prominent foreground objects, as mentioned in Chapter 5, first emerged in *Famous Views of the Sixty-odd Provinces* too. Prominent foreground objects appear in a few prints, as seen from different parts of a ship in "Oki Province," "Nagato Province," and "Iyo Province," as well as the torii gates in "Bizen Province" and "Aki Province." After the successful attempts in the preceding series, exaggerated foreground objects become recurrent in many prints in *100 Famous Views of Edo*.[6]

Famous Views of the Sixty-odd Provinces and *100 Famous Views of Edo* can be seen, in today's pop culture context, as two seasons of an anthology series. Like different seasons of *Street Food* focusing on cities in different regions, the first ukiyo-e print series introduces famous places across 60-something provinces in Japan, and then the second series zooms into Edo and walks the viewer through the famous places in the city. One could imagine if Hokusai had lived a few more years, another sequel might focus on the imperial capital Kyoto.[7] In Chapter 5, the *100 Famous Views of Edo* series is suggested for mapping with the maintenance routine of having takeout lunches. It is commonplace in today's urban lifestyle that a working-class individual finds a place to have a takeout or other quick lunch like those from a convenient store. With recent advances in digital technologies such as mobile devices and AR, one can virtually see different views of Edo from the series, which might depend on his location in the city and the season. As days go by, one might have gone through all famous views of Edo and may start to virtually visit places in other provinces presented by the *Famous Views of the Sixty-odd Provinces* series (or switching the order of the two series, following the chronological order of their publication dates).

Enhancing Continuity across a Series

Routines are typically expected to run for a long time. An analogous story, whether it is a chain tale or theme-based narrative, can be mapped with a routine only for a limited period of time. It is sensible to develop sequels that iterate similar storylines or patterns following the original. A set of sequels to a chain tale form an episodic series in which each episode is a self-contained story engaging the same group of main characters in a similar plot type. A set of sequels to a theme-based narrative constitutes an anthology series in which episodes share a common theme rather than any characters or a plot type. Both episodic and anthology series are able to sustain the themes of the original stories through a balance of consistency and variety. For instance, each episode in "The Turnip and other Root Vegetables" is a chain tale iterating the quest plot type with a new goal, that is, new root vegetable, and an updated team. Mapping episode after episode from the series with a long-running routine continuingly encourages a person to pursue a definite goal. Theme-based narratives, such as *53 Stages of the Tōkaidō* and *69 Stages of the Kisokaidō*, iterate the pattern of traveling on a road connecting Edo and Kyoto. Mapping them with a maintenance routine continuingly adds color to the mundane, repetitive processes.

Yet, both episodic series of chain tales and anthology series of theme-based narratives suffer from one shortcoming—the lack of continuity between successive stories. When a person running regularly to achieve an optimal body mass index has finished "The Turnip," it is sensible for them to start the sequel story, say "The Celeriac." After harvesting a gigantic turnip, however, one might not feel the urge to immediately continue to pursue a new, but similar goal. It is common for one to slow down, take a break, or even procrastinate. The situation can be worse in the case of theme-based stories. When a person has traveled a long journey from Edo and arrived at Kyoto by doing laundry regularly, one feels a sense of satisfaction upon completion, while losing the momentum of starting a new journey right away. Some might argue that it is acceptable to take a break between seasons of a television series. That said, once someone has abandoned an endeavor, they need a reason to pick it up again. Along the same line, people need a reason to start a new analogous story after finishing the first one. Creating continuity between successive stories can motivate a person to continue, but the connections have to be subtle because each story should remain self-contained. Two literary techniques commonly used in narrative may shed light on this issue, namely enveloping a collection of analogous stories in a frame story and weaving an overarching storyline through all analogous stories in the series. The former can be seen as an extradiegetic approach, which adds another layer of unfinished narrative over all stories, while the latter can be intradiegetic by weaving part of a common unfinished narrative into each story. Diegesis, a term originating from ancient Greek, refers to "story." Following Gérard Genette (1980), the extradiegetic is the level above a story, and the intradiegetic level is inside the story.[8] The extradiegetic approach is more suitable for enveloping a set of theme-based stories, because each theme-based story has separate characters and settings, which hinders integration of a common storyline. The intradiegetic approach is more applicable to a series of chain tales, because they share the same set of main characters and largely common settings, which facilitates integration of an overarching storyline. Whether it is extradiegetic or intradiegetic, the loose ends and unresolved mysteries in an "unfinished" storyline are the key motivators for a person to carry on from one analogous story to the next, as informed by the Zeigarnik effect.[9] The theory holds that unfinished tasks are more commonly recalled than completed tasks are, and thus more likely to induce a need to finish them. It follows that a person would be likely to remember the unfinished storyline and develop an urge to continue across successive analogous stories. Even if one chooses to take a break from the routine, the unfinished storyline gives them a reason to resume it sooner or later. In the following sections, let us look at the extradiegetic and then the intradiegetic approaches.

The Extradiegetic Approach

A frame story, simply put, is "a story that contains other stories within it."[10] One can also describe the literary phenomenon of "framing," or reciprocally "embedding," as "a fiction within a fiction"[11] or "a story within a story."[12] The preposition "within" here seems to imply that someone in a story tells another story. In other words, events are narrated by a character in the frame story.[13] Marie-Laure Ryan (1990) sees "frame" as a metaphor that transfers the idea of surrounding visual elements in an image from the visual domain to enclosing a slice of time in the temporal domain. A story frames another story when the verbal representation of the former both precedes and follows the verbal representation of the latter. The phenomenon can be formalized using a system of parentheses, for example, [C1[S1] C1[S2] C1[S3] ...], where S1, S2, and S3 are separate stories narrated by C1, a character in the frame story. This simple example can be comparable to the well-known Arabian novel *1001 Nights* wherein the protagonist, C1 in the example, tells her king husband one tale every night in hope of keeping her from being executed. Meanwhile, some tales told by the protagonist in turn framed some other stories, such as "The Three Ladies of Baghdad." In fact, multiple levels of frames can be seen in other classical novels like Mary Shelley's *Frankenstein* (1818) or Cao Xueqin's *Dream of the Red Chamber* (ca. 1791). In *Frankenstein*, Robert Walton's journal includes a transcript of Victor Frankenstein's narration, which in turn incorporates an oral narration by the Creature Victor has created. The Creature's narration mentioned the history of the De Lacey family, which was yet another level of narrative. In terms of the parenthesis system, *Frankenstein* shows a nested case, looking like [Walton[Frankenstein[Creature[De Lacey]]]]. Other than narration or transcription, another vehicle to transport readers through different levels of frames can be reincarnation or dreaming. The movie *Inception* (2010) is probably the most prominent example featuring many layers of frames through a dream within a dream. Regarding this multiple-level framing phenomenon, some scholars refer to it as "layering."[14]

Two Frame Story Examples

Instead of exploring the depth of framing, some creative works expand the breath of it, using a frame story to sustain readers' interest over a collection of stories, as seen in novels like *The Decameron* (1353) and *The Canterbury Tales* (1387–1400). In *The Decameron*, the frame story is built on a storytelling "contract" among a group of young people, including seven women and three men. They flee the epidemic in Florence and hide themselves in a remote villa. To kill time, they take turns to tell stories. Each day, one member sets a theme, and each of the other members tells one short story accordingly, resulting in ten stories under a common theme. After ten days, they have collected totally 100 stories grouped into different themes. Each story told by each member on a day is self-contained, like each episode in an anthology series. There is no continuity from one story to the next. When a reader reaches the end of one short story told by one member, the reader might feel satisfied and take a break. Yet, the reader remembers that the next member has prepared to tell another story under the same theme. The unfinished list gives the reader a reason to keep on reading. Even though one has finished all stories prepared by all members on a day, it is not the end of the frame story. The reader knows that there are another ten stories on a different theme on the following day. This imminent variety creates anticipation for one to move on. One might also want to know what would happen to the ten young people. Would they survive the epidemic at last? This suspense in the frame story also keeps the reader interested.

In *The Canterbury Tales*, the frame story revolves around a storytelling contest along a pilgrimage. A group of 30 pilgrims travel together from London to Canterbury to visit the shrine of Saint Thomas Becket. They agree to engage in a storytelling contest on the journey. The group consists of many different archetypes, including a knight, a prioress, a physician, a clerk, a miller, a merchant, and others. They take turns to tell one tale for the contest. Between the tales, there are "links" like conversations with the host of an inn or other pilgrims, as well as introductions to the next contestant. Unfortunately, the author Chaucer did not finish the novel in his lifetime, and not all of the 30 pilgrims tell their tales in the book. Imagine a reader having finished one tale in the novel. They feel a sense of completion. Meanwhile, they know that the contest is still not yet finished in the frame story and may want to see how other pilgrims comment on the tale. As the host introduced the next contestant, one would also be curious what kind of tale might unfold next. The unfinished contest stays in the reader's mind and remains a loose end untied forever.

Adding Frame Story to Anthology Series

The frame story is a literary device commonly used to envelope a set of self-contained stories. The stories may share a common theme, like each of the ten short stories told by the ten young people on the same day in *The Decameron*. The frame story is added as another layer of narrative over all the stories. This extradiegetic approach can be applied to enveloping a collection of theme-based stories that are blended with a target routine. Consider mapping the ukiyo-e travel narratives of *53 Stages of the Tōkaidō* and *69 Stages of the Kisokaidō* with the routine of doing the laundry, as suggested in Chapter 5. When a person folds clean and dry laundry at home, they virtually see the ukiyo-e print of the next station on the Tōkaidō, which prompts the imagination of continuing the travel journey to that station. If the person has reached the last station at the end of the Tōkaidō, one needs a reason for starting a new journey on yet another road connecting the same locations. What kind of frame story should we develop to provide the reason? Can the frame story be about an individual who is unable to travel due to some reason (e.g., health condition or work schedule) and so they read travel stories for entertainment or reference until the actual travel becomes possible (as inspired by *The Decameron*)? The framed narratives then include the visual travel stories on the Tōkaidō and on the Nakasendō (aka Kisokaidō). As the framed narratives are blended with the laundry routine, so is the frame story. In the blend of the framed narratives with the routine, the person doing the laundry at home becomes a traveler doing the laundry in a guesthouse. In the blend of the frame story, the person could be an individual waiting for the next trip (like the young people in *The Decameron* waiting for the epidemic to end). Before the next trip, the person reads a visual travel story where they are transported to the Tōkaidō, staying at each station, doing the laundry, and moving on to the next station, until they reach the destination. After finishing the Tōkaidō journey, the person is mentally transported back to the frame story and realizes that the real trip has not begun yet. The person supposes that, like the storytelling contract in *The Decameron*, they have to go through a definite number of travel stories, say ten, before they are able to travel. The outstanding stories that person has agreed to cover are the reason for starting another framed visual travel story about the Nakasendō.

Regarding another pair of ukiyo-e narratives, *Famous Views of the Sixty-odd Provinces* and *100 Famous Views of Edo*, which are suggested in Chapter 5 for blending with the routine of having a takeout or any quick and convenient lunch on a workday. When a working-class person finds a place to eat a takeout, they virtually see the ukiyo-e print of a famous view of Japan from the Edo period, according to one's relative geographical location in the present-day city. The person could imagine having the takeout while traveling to a famous attraction back to the period. If the person has gone through all the attractions of the 60-plus provinces, they achieve a sense of completion. Without any other unfinished task, one does not feel the urge to start another collection of famous Japanese prints. A frame story could provide some unfinished tasks though. Here the frame story can be about an individual who is the host of a visual travelogue competition introducing the entries one by one (as inspired by *The Canterbury Tales*). The entries are the framed narratives, including the two ukiyo-e narratives presenting famous views. In the blend of one framed visual narrative, *Famous Views of the Sixty-odd Provinces*, with the routine of having a takeout, the working-class person becomes a traveler going to various Japanese provinces in the Edo period. In the blend of the frame story, the person could be the host of the competition. While having a takeout on a workday, the person captures virtually a famous view of a Japanese province. When they finish the last province on the list, the person is mentally transported back to the frame story and sees the complete collection of famous views of all provinces. From the host's perspective, one might appreciate or subtly comment on this collection. The host of course knows that it is just one entry to the competition, and there are others not yet being introduced. The unfinished competition reminds the person of the next entry—the next framed narrative covering the famous views of the Edo city that in turn calls for their participation again.

The frame stories of *The Decameron* and *The Canterbury Tales* shed light on how to develop a frame story enveloping a collection of theme-based analogous stories that are blended with a target routine. The frame story needs to be blended with the routine as well. The person practicing the routine definitely takes an imaginary role in the blend between a framed analogous story and the routine. For example, in the blend of the framed Tōkaidō travel story and doing the laundry at home, the person imagines being a traveler on the road. In the corresponding frame story, the person imagines in the blend being someone who entertains themselves by reading travel stories while waiting for the actual travel to happen. On both levels of the narrative, the person is transported into different story worlds, one traveling on the road, and the other likely staying at home reading. Hence, it is extradiegetic. Meanwhile, the person maintains across the narrative levels a consistent interest in travel, that is, the main theme of the analogous stories. In the framed narratives, the person plays the role of a traveler documenting road scenes on the journey. In the frame story, the person plays the role of an audience appreciating each travelogue. The two roles are coherent and may overlap, because one can read one's own travelogue to relive moments in a journey. This coherent role identity makes it easier for the person to mentally traverse different story worlds on different narrative levels while performing a routine. This case can be called coherent-identity extradiegetic approach. The external knowledge frame invoked for this blend can be [reading]. New relation to be imported can be READ (travelogue, journey), which means reading a travelogue about a journey. Together with the already imported relations in the blends of the framed narratives, including CHECKIN (guesthouse) and WASH (clothes, guesthouse), the person staying at home can imagine a third story of reading their own travelogue while waiting for the washing and drying cycles to finish.

On the other hand, in the case of the framed narratives of famous views and the corresponding frame story of a competition, as inspired by *The Canterbury Tales*, the person practicing the routine of having a takeout lunch seems not only landing on different story worlds on different narrative levels but also taking opposing stances or roles. On the framed narrative level, the person imagines being a traveler during the Edo period. On the frame story level, the person turns themselves into the host of a competition that may happen at any point of time. The person changes their role with different identities. In the framed narratives, they play the role of a traveler capturing famous views at different places for a travelogue. In the frame story, they play the role of the host introducing each visual travelogue. The two roles, although both representing someone who are interested in famous views, take different stances in a competition. A competition host typically cannot simultaneously be a contender in the competition. It requires more mental effort for the person to shift their perspective between the two contrasting identities. This case can be called double-identity extradiegetic approach. The external knowledge frame for this blend is [contest], and the new relation to be imported should be INTRODUCE (host, entry), which means the host introducing each entry. The person can imagine a third story of having a takeout while hosting a competition and introducing the next entry, which then transports them to the next journey of capturing famous views.

Whether it is the coherent-identity or double-identity case, the changing role that the person plays on different narrative levels still maintains the same interest in the main theme of the analogous stories. The frame metaphor in literature is built on the aesthetics of framing a picture. The stories inside the frame are separate from the story outside of it. Like a picture frame, the frame story holds different framed stories together. While a frame story of a literary work "surrounds" other stories on a linear, verbal dimension, some intriguing creative works metaphorically and literally frame their substances on a spatial, visual dimension. The framing approach provides a useful reference for mapping and blending with routines as well. In the next section, I will introduce an intriguing traditional artifact from the imperial China.

A Frame Story: Counting Days using the Nine-Nines Charts

The nine-nines charts, or counting the nines charts, are traditional visual artifacts functioning like a special calendar for people in ancient China to mark and count the days from the winter solstice to the next spring. The winter solstice is the day with the shortest hours of daylight and the longest night in a year. In the Northern Hemisphere, it typically falls on the 21st or 22nd of December in the Western calendar. In the traditional Chinese calendar, the day, which is called "Dongzhi" (冬至) literally meaning "the extreme of winter," occurs in approximately the second last month of the year. After Dongzhi, hours of daylight become increasingly longer by the day, resulting in gradually warmer weather. After the Lunar New Year, comes the next spring. Typically, there are about three months between Dongzhi and the Beginning of Spring. In the old days, the Chinese people have developed intriguing visual means of marking and counting the days of this bitterly cold yet hopeful period. They divided it into nine nine-day periods. The number "nine" is arguably linked to the "three" yin and "three" yang lines in the eleventh hexagram Tài (泰) of the *I Ching*, or Book of Changes, which represents "auspiciousness."[15] They created varied forms of nine-nines charts, each of which consists of nine clusters of nine entries together with other visuals. The earliest textual reference of the nine-nines charts can be dated back to the Yuan dynasty (ca. 1271–1368). The poet Yang Yunfu mentioned a nine-nines counting chart in his work. Maggie Bickford (2005) translated the texts as follows:

> Try counting off the window's nine-nines chart,
> Then lingering cold has all run out and warmth's returning starts.
> [When] the plum-blossom dots show no trace of white spots,
> What you see this morning is flowering apricots.

A note following the poem further explains how to use the nine-nines chart. After Dongzhi, a branch of plum is pasted on a paper window screen. A lady puts on makeup in the morning and uses rouge to draw a circle on the paper every day. When 81 circles are drawn, the plum blossoms (typically in January) are spent and replaced by apricot blossoms (from February to March). Plums and apricots have similar flowers, but apricots blossom later. The apricot flowers indicate the return of warmer weather. This act of drawing circles is thus called "counting the nines." While there exist more references in literature from the Yuan and then Ming dynasties, the earliest visual evidence should be an ink rubbing dated 1488. The center of the image shows a plum spray in a vase, bearing nine clusters each comprising nine flower petals. The image has a clear title, "Jiujiu xiaohan zhi tu" (九九消寒之圖), literally meaning "chart of nine-nines disperse the cold." Its visual structure, with two layers of nines, seems to be the fundamental structure of nine-nines charts.

Built on the two-layer structure, the nine-nines charts have evolved into new forms during the Qing dynasty (1636–1912). Petals are transformed into symbols engaging meaning, such as Chinese characters constituting a poem. Some of these nine-nines poetry charts still survive today. A particular chart is still mounted on a wall inside the Forbidden City of today (Palace Museum in Beijing). The chart is a rectangular frame (about A2 size) presenting nine Chinese characters, each of which is constructed in nine strokes, in a three-by-three matrix. The nine characters, starting from the top-right corner, top to bottom, then right to left, form a poem (亭前垂柳珍重待春風), which can be translated as "in front of a pavilion, weeping willows, take care, and waiting for the spring breeze." The chart is titled "Guan cheng chun man" (管城春滿) informing that "brush strokes complete the spring." The original calligraphy of the poem, which is said to be attributed to the Daoguang emperor (reigning 1820–1850), was copied in double outlines and then printed for distribution in that period. Similar charts were mass-produced for the market too.[16] One could imagine that a person used the chart to count the days from the winter solstice to the first day of spring. The person started with the first character in the top-right cell of the matrix, writing one stroke each day. This character (亭) is also a pictograph, visually resembling its referent, "pavilion." As strokes accumulated day after day, the character and its image gradually appeared in the cell (that is also a frame). After the first nine days, brush strokes completed the image of a pavilion in the top-right frame.[17] This was one first-level framed narrative enabled by this chart (see Figure 6.1 for a simulated sequence of writing the strokes on the chart). Writing nine characters one by one and counting 81 days, the person went through nine first-level framed narratives, each of which presents a sequence of strokes forming an image in a cell (i.e., each small frame).

After finishing the nine characters, the person finally saw them all as a whole (see the final look of the chart in Figure 1.3). One could read them in sequence.

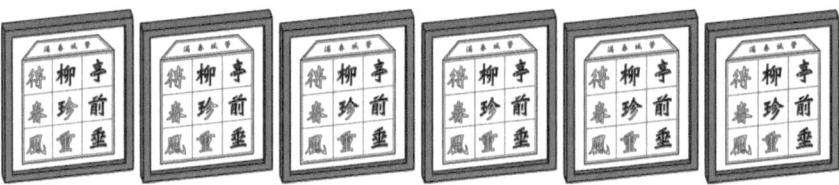

FIGURE 6.1   A simulated sequence (from left to right) about writing the strokes on a nine-nines chart in the Daoguang period.

FIGURE 6.2   Another remake of a nine-nines chart in Puyi period, with part of the weather descriptions made on the first and fourth strokes of the top-left character, and the last stroke of the center character. Calligraphy by Shum Sheung Yin.

They form a poem written within the overall frame of the chart. Poetry evokes images and emotions. The person could visualize a scene in which people might part in winter while looking forward to their reunion in spring. One could sense the poignant yet hopeful moment. This is the second-level framed narrative. Finally, there is a frame story. It is about the person who counted the days and documented the world where they inhabited. The person's primary goal was to count the days through the bitter winter to usher in the flourishing spring. Through marking the days since Dongzhi, one saw how many days to go before the advent of spring. The visual representation of the chart, whether it is a spray of plum blossoms, a set of abstract symbols, or a sequence of Chinese characters, allowed a direct perception of progress made toward the next spring at a glance. The ratio of written brushstrokes to those empty strokes to be filled provided an immediate, tangible progress indicator. Seeing the brushstrokes gradually accumulating day by day, one felt hopeful toward attaining a goal, which satisfied one basic psychological need, that is, competence. The nine-nines charts narrate analogous stories that strike a balance between variation and progression.

Meanwhile, people might use the chart to record the weather of each day too. Another chart surviving in the Palace shows evidence of this practice (see Figure 6.2). This chart is believed to be created in the final year (1923/1924) of the last Qing emperor Puyi staying in the Forbidden City. The person in the palace, after writing a stroke, jotted down a brief description of that day's weather in white on top of the black stroke. In the surviving sample chart, the first nine seem to start on the sixteenth of the eleventh month that year, as noted on the bottom-left corner of the first cell (top-left) of the overall frame. The first character in this chart is "Chun" (春) meaning "spring," and the first stroke is the topmost horizontal one (一). One top of it, the small characters in white say "sunny, cold day." The fourth stroke should be the long one from the top of the character to its left (丿). The description on it says "chilly wind hits the body like a knife all day." At the center of the matrix, the character is "Feng" (風) meaning "wind," and the last stroke of this character is the bottom-most dot (丶). It corresponds to the first day of the Lunar New Year, with the description "chilly" on it. When the person finished this stroke, they saw a complete "Feng" character in the cell. One felt satisfied with the completion of the fifth character and appreciated the writing of the poem. Meanwhile, the small white characters reminded the person of taking note of the weather. They immediately jumped from the framed narratives back to the frame story, resuming the role to count how many days had passed and to record the day's weather. The intriguing part is how the changing weather documented by the person resonated with the semantic content of poem. The poem in this particular chart (春前庭柏風送香盈室) can be translated as "before spring, cypress in the courtyard, breeze bringing scent to fill the room." Like the other chart, this poem projects a sensory and pleasurable image of spring. This framed narrative shares a common theme with the frame story of counting and documenting the days from winter to spring. The main theme of the analogous stories here is the auspicious spring. The nine-nines poetry charts demonstrate the extradiegetic approach.

Suggested Routines

The above nine-nines poetry charts contain analogous theme-based stories that are already blended with the routine of marking and counting the days over a period. One could also blend a nine-nines chart with other daily or weekly maintenance routines, such as regular self-monitoring of blood glucose level, applying facial cream or cosmetics every day, or walking the dog for 30 minutes a day. Consider a person who wants to keep walking their dog every day in winter and uses the chart featuring the poem starting with "Chun" (春). The character is based on the simplified hieroglyphs of grass and the sun. In the first nine-day period, the person might choose nearby places with a lawn (e.g., a park) to walk their dog. After each day's walk, they write a stroke, which can be a virtual stroke simulated by a system running on a tablet computer. Day by day, the character "Chun" gradually builds up. This is like the person capturing what is seen (i.e., lawn) during the walk and recording their experience bit by bit. After nine days, the visual record is completed. The second character "Qian" (前) is an ideograph, consisting of a foot and a shoe, which indicate a forward movement. The person can take any path with the dog in the second nine-day period. The third character "Ting" (庭) means courtyard. The person can find a path passing by a courtyard of some sorts. Hence, for every nine-day period, the person repeatedly walks the dog on a path with some view alluding to the corresponding character. Every walk reinforces one's mental image of the path and adds extra to the record. It takes nine days to complete one character. Repeating the routine nine times, one results in nine characters forming a complete poem, which tells the framed narrative.

Imported Relations for Blending

After walking the dog on a path with lawn nine times, the first character "Chun" appears in the chart. The pictographic character can be seen like a visual record of the scenery along the path. This evokes an external knowledge frame (parallel to the narrative frame) [visual journal], which is a way of recording life using visual media in addition to text. A relation that can be imported is SKETCH (thing seen, line, day), which means the act of sketching that turns what one has seen on a day into a line drawing. For the first character, the relation binds together imagery of grass on a lawn one has seen and the brushstrokes. When strokes accumulate after each walk, the person can imagine a third story of going out each day, remembering the outdoor scene, sketching part of it on paper after returning home, and completing a line drawing in nine days. This is a blend of the chart's first-level framed narrative with the routine. Another relation imported from [visual journal] for the blend of the second-level framed narrative can be CAPTION (illustration, text). Caption is a piece of text typically appended to an illustration as a brief explanation or title. It can also be like a tag informally. When all the nine characters have appeared to form a poem, it documents the senses of touch (through the breeze) and smell (from the cypress) in parallel with the visuals in the moment. The poem can be seen as a tag in words describing the pictures embedded in the characters. The person can imagine another third story of writing a poem to append to the illustrations in the visual journal after practicing the routine for a period of 81 days. That is a blend of the second-level framed narrative with the routine.

When the poem is completed in the chart, all framed narratives are concluded. It seems like the frame story, which is about someone counting the days from winter to spring, also drawing to a close. Yet, this is not a close but a hiatus. The four seasons repeat every year. The agonizing winter would come again. The frame story would continue in the next winter with an updated poem and characters. This is the reason why there are different versions of the nine-nines poetry charts. The version featuring the poem starting with "Ting" (亭前垂柳珍重待春風) was believed to be created during the period of Daoguang, prior to the one starting with "Chun" (春前庭柏風送香盈室), which appeared at a time when Puyi was active. The latter can be regarded as a sequel to the former. As said, these charts have been in circulation outside the Palace. Other poems with the same nine-nines structure can be found in the folk culture, such as one starting with "Yan" (雁南飛柳芽茂便是春) that can

TABLE 6.6 The formal representation of the blends between walking the dog and the nine-nines poetry charts

Walking the dog	Selective projection & pattern completion	Counting days on nine-nines poetry charts
	[visual journal]:	
WALK (person, dog, Day1) →	WALK (person, dog, Day1)	
	→ SEE (person, grass)	
SEE (person, grass)	SKETCH (grass, 一, Day1) ←	WRITE (一, Day1)
WALK (person, dog, Day2) →	WALK (person, dog, Day2)	
	→ SEE (person, grass)	
SEE (person, grass)	SKETCH (grass, 一, Day2) ←	WRITE (一, Day2)
WALK (person, dog, Day3) →	WALK (person, dog, Day3)	
	→ SEE (person, grass)	
SEE (person, grass)	SKETCH (grass, 一, Day3) ←	WRITE (一, Day3)
WALK (person, dog, Day4) →	WALK (person, dog, Day4)	
	→ SEE (person, grass)	
SEE (person, grass)	SKETCH (grass, 丿, Day4) ←	WRITE (丿, Day4)
…	…	…
	COMPLETE (1st picture of grass and the sun, nine days)	COMPLETE (春, nine days)
	…	
	[visual journal]:	
WALK (person, dog, Day10)	COMPLETE (2nd picture, nine days) ←	COMPLETE (前, nine days)
…	…	
WALK (person, dog, Day19)	COMPLETE (3rd picture, nine days) ←	COMPLETE (庭, nine days)
…	…	
WALK (person, dog, Day28)	COMPLETE (4th picture, nine days) ←	COMPLETE (柏, nine days)
…	…	
WALK (person, dog, Day73)	COMPLETE (9th picture, nine days) ←	COMPLETE (室, nine days)
…	COMPLETE (poem, 81 days) ←	COMPLETE (poem, 81 days)
	CAPTION (pictures, poem)	
	[calendar]:	
	COMPLETE (winter, 81 days) ←	COMPLETE (winter, 81 days)
	REMARK (poem, person, dog, 81 days)	
	FOLLOW (winter, spring)	
…	…	…

be translated into "Geese fly south, willow buds flourish, that is spring." Another poem starts with "Xing" (幸保幽姿珍重春風面), which can be translated as "Fortunately preserve the elegance, take care of the face like spring breeze."[18] These poems can be put together as an anthology series of nine-nines poetry charts. Each of them projects a different imagery between winter and spring. Yet, they all share a common theme—hopefully looking forward to the auspicious spring. Enveloping all these poems, the frame story of counting the days and documenting the weather continues every winter. Regarding the person walking the dog, the routine should last extensively across seasons. The reason for continuing can be found in the blend of the frame story. After finishing the poem "Ting" in a winter, one also completes the visual journal and feels satisfied with the sketches and the caption. Meanwhile, the person recognizes that the chart is indeed a winter calendar, which is followed by a spring calendar. The person could continue charting with the poem "Chun" or "Xing," whose imagery and emotion match the mood of spring better. Replacing a completed poetry chart by another new one is like turning a new page in a modern calendar. An external knowledge frame [calendar] is then invoked, and new relation FOLLOW (winter, spring) is imported for blending walking the dog with the frame story of counting and marking the days. Both the routine and the frame story continue season after season. In this blend, the person can imagine a third story of replacing a new calendar for the new season and continuing to walk their dog every day. Table 6.6 summarizes the above blending processes.

The Intradiegetic Approach

After concluding an analogous story blended with a routine, a person might want to take a break and later would need a reason for starting a new story and continuing the routine. According to the Zeigarnik effect, an unfinished storyline would make the person remember it and develop an urge to resume the action after a break. Adding a frame story above all analogous stories is an extradiegetic way of keeping one storyline unfinished. This approach works well for an anthology series of theme-based stories. Another approach is weaving the unfinished storyline through all stories. This intradiegetic way is more applicable to an episodic series of chain tales, because they share the same set of main characters and largely common settings, which facilitate the integration of an overarching storyline. In fact, the intradiegetic approach to building continuity across episodes is a common technique in series or serials in varied media.

Series and Serials

Before moving on, let us differentiate between series and serials. Technically, there is a clear distinction between series and serials. As mentioned, an episodic series consists of self-contained episodes. Although each episode engages the same set of main characters, storylines do not continue across episodes. In an anthology series, episodes even have separate stories of completely different characters and settings. A serial, on the other hand, is "a continuing narrative distributed in installments over time."[19] Each installment is not concluded until the end of the series.[20] The origin of the serial form can be dated back to the nineteenth century, when the works of authors like Charles Dickens were published serially in periodicals prior to being published as novels. This practice caught on across the globe in the later half of the last century. For example, in East Asian context, Chinese martial arts and warriors fiction, which is called "wuxia," a popular genre by the times, was often initially released as installments on newspapers or magazines, such as the epic works of Jin Yong (金庸) published from 1955 to 1972, and the more character-focused series of novels by Gu Long (古龍). Each of Jin Yong's works is a self-contained exceedingly long novel involving a specific set of main characters situated in a particular period of the imperial China, while Gu Long is famous for his series of novels which tell different stories revolving around the same protagonist. Of these wuxia novels, the initial publications are in the serial form. For decades, Japanese comics (aka manga) have been first released as installments, typically a dozen pages each, in thick, weekly, or monthly anthology manga magazines, which often have hundreds of newsprint pages in each issue and many individual series by different authors. Multiple installments of a single series are then collected and reprinted on higher quality paper as a paperback-sized volume, which is called "tankōbon" in Japanese. Popular works typically can last for dozens of volumes, such as *Doraemon* (1969–1996), *Crayon Shin-chan* (1990–2009), *Dragon Ball* (1984–1995), *One Piece* (1997–present), and *JoJo's Bizarre Adventure* (1987–present). Some of these works, like *Doraemon* and *Crayon Shin-chan*, are typical episodic series, featuring the same set of main characters involved in numerous separate, self-contained short stories. Yet, other works have much longer, continuing storylines. Some of them may be divided into parts (e.g., *JoJo's Bizarre Adventure* has nine parts so far), but all parts continue on the same timeline and characters develop. Although these comics are sometimes called series, the narratives are distributed in the serial form. In fact, many long-running television series and comic series are combinations of both episodic series and continuing serials, what Robin Nelson (1997) calls "flexi-narrative" (30–49). Some series literally consist of some episodes that are self-contained and others that contributes to the large story arc. For example, most episodes in a season of *The X-Files* (American TV series, 1993–2002) deal with "stand-alone," "one-off" cases, while a few episodes, usually toward the end of the season, follow the long-standing quest to uncover the conspiracy of alien existence and their invasion plan. Other series, on the other hand, interweave both self-contained and continuing storylines in each episode. The self-contained storylines start and end within the episode, while some events actually advance the overarching narrative, or connect to it in a subtle way that is realized in later episodes. For instance, in *Sherlock* (British television series, 2010–2017) the protagonists mainly focus on resolving one crime case in each episode, while Sherlock Holmes subtly picks up minor clues in each episode that add up to reveal the mastermind behind all cases. The second approach that combines the variation of episodic series with the progression of continuing serials informs the intradiegetic approach here.

Some media theorists are interested in further characterizing narratives told in the serial form. Among the characteristics, iteration seems to be a major aspect. Frank Kelleter (2017) proposes "five ways of looking at popular seriality," highlighting aspects including evolution and proliferation, which to a certain extent are associated with repetition and variation of serial narratives resulted from the continuing interactions between creation and reception of the serial forms. Serial narratives are segmented into installments that are released one by one from time to time, say weekly or monthly. The time lapses between installments allow production schedules overlapping with that of publication or distribution. That means, a serial narrative is continuingly being written while its early part is already being read or watched. Serial producers may respond to the reception of early installments and adjust how it continues in later ones. Hence, serial narratives are said to be evolving in a feedback loop between production and consumption. Well-received elements may be repeated and elaborated; unexciting parts are left behind and replaced by other options in later installments. This kind of repetition and variation often shape the development of supporting characters or antagonists. Consider the American superhero comics like Batman. Among the countless villains since its debut in 1939, quite some of them recur in different issues, and a bunch of them finally reach the status of so-called "supervillain," such as Joker, Penguin, and Riddler. A few supporting characters have remained as long-time partners of Batman, like Robin and Catwoman. Some of them have their own spin-off series and extended characters, such as Joker and his partner Harley Quinn. The evolution and proliferation are the results of a long-time "natural selection" process involving both the producers and consumers in an iterative feedback loop running across decades of installments.

Iteration and Multiplicity

The nature of iteration is explicitly named as one of the six elements of serial narratives by Sean O'Sullivan (2019). The first three elements are described as patterns within installments, namely, iteration, multiplicity, and momentum, which are all related to the discussion of the intradiegetic approach here. By iteration, O'Sullivan means that a serial narrative always repeats a certain component across installments, which becomes its "definitional, recognizable feature" (Sean O'Sullivan 2019). One prominent example is the so-called "cold open," or pre-title sequence, in many television series, such as *Six Feet Under* (American TV series, 2001–2005) or *Bones* (2005–2017). Every episode of *Six Feet Under* starts with a short sequence showing how someone died in an unusual way, which leads to a funeral case resonating with the main theme of the series. In case of *Bones*, the cold open sequence is always an unexpected discovery of a dead body, or parts of it. Another example is a common pattern in many Japanese anime or tokusatsu (live-action with special effects) television series, a key sequence repeating every time when the protagonist applies the supernatural power. In *Sailor Moon* (Japanese anime TV series, 1992–1997), when the teenage female protagonist uses her ability to transform into the magical Sailor Moon to battle the villains in each episode, the same sequence of shots showing appearance of her suit and accessories is inserted. Similar sequence can be traced back to the classical tokusatsu television series *Kamen Rider* in 1971. The protagonist can use the belt with mysterious technology to transform himself into the full-fledged mode of a grasshopper mutant. These iterative, signature sequences featured in serial installments are comparable to the repeating actions (e.g., pulling up a root vegetable) in the analogous chain tales that are blended with routines.

While the first element of seriality made by O'Sullivan, iteration, focuses more on repetition, the second element, multiplicity, relates to variation. Variation in serial narratives can be seen in terms of one-off villains, characters, and backstories, as well as cases. In *Sailor Moon* or *Kamen Rider*, villains in each episode look different, although all of them are defeated by the superhero using similar attacks. In *The X-Files* or *Sherlock*, the one-off case in each episode comes with completely different characters and backstories, while they all seemingly or ambiguously connect to the large story arc. O'Sullivan uses one of the first American film serials in the silent era, *The Perils of Pauline* (1914), to illustrate his point. The serial consists of 20 episodes of 20–30 minutes each, released in bi-weekly installments. The story is about an ambitious and independent young heiress, Pauline, in pursuit of different adventures before settling down in a marriage. The villain is her father's secretary who intends to get rid of her to receive the inheritance. The episodes thus present Pauline's series of adventures, which always cause Pauline's life to hang by a thread because of the villain's cunning schemes. Each episode ends with the problem solved and the protagonist saved. The plot of each adventure is quite iterative, as O'Sullivan puts it, but the genre shifts from treasure hunting, traveling to the West, to crime drama. The main activity in each adventure also changes from a balloon ride, flying an airplane, driving a racing car, or acting in a movie, to a submarine tour. While the repetitive pattern in the plots maintains the serial's motif, the variation in genre and activity seems to be intended for creating a sense of novelty across the episodes and attracting a broader audience.

Momentum: A Continuing Quest

The third element, momentum, is the most relevant to the intradiegetic approach to enhancing continuity across an episodic series of analogous chain tales blended with a routine. Momentum is the energy or force that keeps something moving, which, in the context of serial storytelling, refers to the part of the story that keeps one reading or watching one episode after another. The cliffhanger, putting the protagonist in an unresolved, critical situation to create uncertainty or suspense, is an overly used common technique, and it is not applicable to episodic series where loose ends are tied and mysteries are solved immediately in each episode. Interweaving an overarching storyline into each episode is a good way of keeping the momentum. That is, each episode has a small narrative that begins and ends within the episode, while advancing another large narrative or incrementally revealing details that lead the characters to conclude the large narrative after a number of episodes. The large story arc can be a continuing quest. The film serial *The Perils of Pauline* again demonstrates an example of building the story arc on the protagonist's internal quest. Pauline, the protagonist, is an adventurous woman and is interested in writing about her own adventures. But such disposition and pursuit are in conflict with the marriage proposal that she has received. She thus decides to fulfill her own desire first before accepting the proposal. This set the story arc. After each adventure, audience might be curious about how the protagonist feels. Does she think that she has had enough adventures? Does she feel that it was time to settle down? When she said that she was not ready yet, the inconclusive proposal remained a suspense in the large narrative extending to the next likely perilous adventure.

A continuing quest built on the protagonist's desire can be a narrative arc integrated into a series of analogous stories of the quest plot type, like "The Turnip and other Root Vegetables." The series include iterative, analogous chain tales where the pleasant family harvests different gigantic root vegetables from their field, namely, a turnip in winter, a celeriac in spring, a carrot in summer, and a beet in autumn. The grandfather might be thinking of retirement and selling his farm to support his granddaughter's education. However, he wants to win the championship once in the biggest root vegetable contest before retiring. This internal quest leads to the overarching storyline. After pulling out the first gigantic turnip, the family submits it to the competition but without winning any prize. The person engaging in a workout routine may wonder about the grandfather's next plan. Would the grandfather decide to sell the farm and retire? Knowing that the grandfather wants to farm for another season, the person continues with the quest with another kind of root vegetable. A goal not yet reached in the large narrative is like an unfinished business on the person's mind, reminding him to carry on with the routine and wait for the next harvest. For showing progress in the story arc, the gigantic celeriac might earn the peasant family the third prize in the contest. But the grandfather still wants the championship. The incremental advancements provide the person with the momentum to move on to the next chain tales in the series. This overarching, continuing quest to be the champion can be blended with the person's workout routine. An external knowledge frame [self-determined goal] comes with a new relation SELF-ASSESS (performance, goal), which compares one's own performance with a self-determined goal. This relation can be imported to bind together the grandfather's act of continuing to farm in pursuit of a larger root vegetable and the person's self-review of the progress made toward the goal. The person can imagine a third story of continuing the pursuit together with the peasant grandfather.

Momentum: An Enduring Monster

Other than a continuing quest, the narrative arc of a series of iterative and stand-alone stories can also be of another plot type, that is, overcoming an enduring monster. Built on "overcoming the monster" plot, the protagonist has to deal with a series of threats coming from the dark force, which is hard to eradicate. In *The Perils of Pauline*, the dark force is the main villain, that is, the wicked secretary who wants Pauline's inheritance. This villain repeatedly and relentlessly puts Pauline in perils, until the end of the serial. Yet, Pauline has no intention to take down the villain. She just wants to continue with her unfinished quest. In this sense, the Japanese tokusatsu television series *Kamen Rider* features a story arc that is obviously overcoming an enduring monster. The superhero protagonist in the series, Kamen Rider, directly confronted the dark force, called "Shocker," which is a mysterious terrorist organization. Shocker used technology to transform and brainwash innocent people into mutant cyborgs to advance their evil plans. The protagonist is one of the victims who was once transformed, but he escaped with his conscience and mind intact. He aims to wipe out Shocker, while fighting and defeating other brainwashed mutant cyborgs in every episode. It took 98 episodes from 1971 to 1973 for the superhero to (seemingly) overthrow the evil organization. While watching the series, the audience would enjoy the battles between Kamen Rider and varied, peculiar forms of mutant cyborgs. After seeing the superhero resolving an immediate threat at the end of each episode, one would feel satisfied but also know that Shocker will return. One would also wonder who is behind Shocker or if Kamen Rider is getting closer to beating the ultimate leader after each fight. The "enduring monster" in the story arc maintains the suspense and symbolizes the unfinished business.

Overcoming an enduring monster can be considered for overarching a series of analogous chain tales about "overcoming the monster," like "The Three Little Pigs and other Threats." In the series, the little pigs iteratively face attacks from different predictors, first a wolf in spring, and then a bear between spring and summer, followed by three wolves in late summer, and finally a polar bear before winter. Every time, the little pigs defend themselves and manage to survive the threat unscathed. One could imagine a main villain behind all these attacks. This villain's ill intent should be compatible with the person's goal behind the routine, in this case, taking medications or supplements to boost their immune system. The villain can be a human real estate developer wanting to get rid of all residents in the area, including the little pigs, for redevelopment. This malevolent developer directs different predators to attack the little pigs. After seeing the little pigs successfully get rid of the first wolf, the person practicing the routine might feel slightly relieved, while assuming that the medications are effective. Yet, they also know that the villain will strike back and so vigilantly stick to the routine. The malevolent developer's evil plan can be blended with the routine. One can invoke an external knowledge frame [defense system] which is applicable to both the enduring monster and the routine, and import a new relation RETURN (attack, host). The relation binds together the return of the malevolent developer to attack the house and the recurrence of an ailment. A third story imagined can be that one has to maintain both a strong immune system and a secure house to brace oneself for the return of the enduring monster.

In popular culture, series and serials exist at the two extremes of a continuum of narrative continuity. Anthology series (e.g., *Black Mirror*) have the least continuity, with separate stories, different characters, and even different contexts between episodes. Episodic series (e.g., *Dorameon, Bones*) see very limited continuity only in a similar set of characters, who may or may not develop across episodes. Serials (e.g., wuxia fiction) show the maximum continuity, distributing continuing narratives in installments. Between episodic series and continuing serials, lie many of today's television, comic, or even film series or serials (e.g., *Sherlock, JoJo's Bizarre Adventure*, Marvel's superhero movies). The terms "series" and "serial" have also appeared in the arts. Starting from Monet's paintings series such as *Haystacks*, some artists have often focused on one subject matter across a series of work driven by a premise. *Haystacks* can be seen as the result of the artist's premise, that is, rendering "instantaneity" in the same light and air.[21] *Haystacks* or other similar series is comparable to an episodic series. The similar subject matter reappears in different pieces, like the same main characters being involved in different stories. In Japan, ukiyo-e print series like *100 Famous Views of Edo* are the outcomes of the ukiyo-e artists' premises. They correspond to anthology series of theme-based narratives, whose continuity is the least. The prints are only related in theme and visual composition, just like the episodes in an anthology series. When the arts continued to evolve in the latter half of the last century, some artists built their premises on systematically predetermined processes and rules, resulting in what can be called "serial art." With the processes and rules strictly followed, a serial artwork often demonstrates repetitions and variations with high internal continuity. Examples included Frank Stella's paintings of "interlaces," "rainbows," and "fans" in the *Protractor Series* (1967–1971) and Sol LeWitt's *Incomplete Open Cubes* (1974/1982). LeWitt's work can be related to the intradiegetic approach to enhancing continuity. I will talk about this in the following section.

A Serial Project: Cataloging Variations in *Incomplete Open Cubes*

In the 1960s, Western contemporary artists started to cast questions on the commodification of art objects. As a result, a few relevant schools of thought emerged, namely, conceptual art, minimalism, and serial art. Conceptual art was first referenced by LeWitt in 1967:[22]

> In conceptual art the idea or concept is the most important aspect of the work. When an artist uses a conceptual form of art, it means that all of the planning and decisions are made beforehand and the execution is a perfunctory affair.

LeWitt, "Paragraphs on Conceptual Art," Artforum Vol. 5, no. 10, Summer 1967, pp. 79–83

In other words, the concept of a work is more important than the finished art object. When it comes to serial art, the concept can be an artist's premise that determines a plan, and the art object is just a log of the process and outcome. In accompanying one of his works, *Serial Project, I (ABCD)* (1966), LeWitt wrote, "[t]he serial artist does not attempt to produce a beautiful or mysterious object but functions merely as a clerk cataloging the results of his premise."[23] To be more specific, the premise is embodied as a systematic process with a set of rules. The artist repeats the process and follows the rules to exhaust variations, making a systematic list of results as an "artwork." Minimalism also plays an important role in many serial artworks. The idea posits that art should not be an imitation or representation of an aspect of the real world such as a landscape or a person, or reflecting an experience or an emotion. Hence, minimalists distill away images or other contents and keep only simple geometric shapes like squares and arcs.[24] For example, Joseph Albers's work relentlessly puts the square at center stage, Stella's *Protractor Series* feature arcs and bands, and LeWitt's projects deal with cubes.

Incomplete Open Cubes, one of LeWitt's projects, is an exemplar that illustrates the iterative nature of serial art.[25] The outcome of the project includes an installation of 122 wireframe cubes, what the artist calls "open cubes," together with drawings and diagrams of the structures.[26] The artist's predetermined plan is to exhaust all possible cubic structures with one or more edges removed. The rules can be summarized as follows:[27]

– The structure should be three-dimensional.
– The structure should be connected.
– Two structures are considered to be identical if one can be transformed into another by a space rotation (but not by a mirror reflection).

A typical cubic structure consists of 12 edges (see Figure 6.3). One could imagine that the artist started with the least incomplete structure by removing one edge and found only one possible 11-edge open cube. Then, he randomly removed one more edge, and put the ten-edge open cube in his log. To find another ten-edge open cube, the artist had to take out two arbitrary edges from a complete open cube and compare it with the other ten-edge open cube in his log. He double-checked his log to avoid duplication by rotating the potentially new incomplete open cube in various directions and angles. This was a tedious and effortful process. The artist repeated the process for every newly found candidate, until he believed that all possibilities had been exhausted. At this stage, he built a catalog of five possible ten-edge open cubes. This was like a conclusion of an episode, supposedly titled "Variations of Ten-edge Open Cubes" (see Figure 6.3). Yet, this was not the end of the overall plan to find out all incomplete open cubes. The artist must move on to the next episode of nine-edge open cubes. He repeated the similar process,

FIGURE 6.3   A complete open cube composed of twelve edges (top) and five variations of ten-edge open cubes (bottom) in isometric projections of three-
dimensional forms, wherein the edges at the back of a cube are partially
shaded to create an illusion of depth.

removing three edges this time and checking repeatedly to avoid duplication. The
list of nine-edge open cubes was longer and it took more time and effort for building. After exhausting all possible nine-edge open cubes, the artist increased the
incompleteness to having four edges removed, that is, eight-edge open cubes. His
quest would go on until the list of three-edge open cubes, the least number of
edges for a three-dimensional structure, was done. The whole serial project can be
seen as a continuing quest, that is, an overarching plot, enumerating all degrees of
incompleteness from the lowest (i.e., having one edge removed) to the highest (i.e.,
only three edges remaining). For a particular degree of incompleteness, the iterative process of finding and cataloging variations constitutes an episodic narrative.

The enumerative component of the serial project is strikingly similar to that in
the aforementioned serial narratives in film or television. In *The Perils of Pauline*,
the protagonist's continuing quest seems to enumerate different adventurous activities, including riding in a balloon, flying an airplane, driving a racing car, riding
in a horse race, going on a treasure hunt, acting in a motion picture, and touring
a submarine. In *Kaman Rider*, the mutant cyborgs appearing in different episodes
enumerate blends with different non-human species, including arthropods like a
spider, a scorpion, a mantis, or a wasp; reptiles like a chameleon, a cobra, or a
gecko; mammals like a bat or a wolf; sea-dwelling species like a piranha, a starfish,
or a crab; birds like a condor; as well as plants like a pitcher plant and a cactus. In
fact, enumeration is a process embodying multiplicity, one element of serial narrative in the O'Sullivan's list. It adds variety to complement repetition in seriality.

Suggested Routines

The enumerative and repetitive nature of serial art renders works like *Incomplete
Open Cubes* appropriate for blending with routines that are driven by beliefs.

The minimalist, non-representational approach of serial artworks also allows more flexibility in mapping with routines. The pure geometric forms let one focus on the processes that correspond to the routine acts. For example, the removing (or adding) of edges in *Incomplete Open Cubes* can be comparable to the act of adding a fixed amount to a pool in belief-driven routines like regular money saving, or more specific, dollar cost averaging. Dollar cost averaging is a generally acknowledged investment strategy, suggesting that a person invests the same amount of money in stocks on a regular basis, for example, every month, every other week, or even every week. In principle, it minimizes risk and lowers the overall cost for the shares purchased over time.[28] With a belief in the strategy, one might engage in on-going investment by, say, setting up a pre-authorized bank transfer into an investment account. Yet, people might prefer flexibility and choose to do it manually. Oftentimes, dollar cost averaging aims at long-term investment that takes months or years to see noticeable return. It requires one's self-determination and perseverance. Blending the routine with analogous stories would motivate one with the benefits of expressive iteration.

Imagine someone who plans to invest $100 per week in stocks. An analogous story for blending with the regular buying routine can be based on a walkthrough of *Incomplete Open Cubes*, starting from the most incomplete cubes (i.e., the least number of edges) toward the least incomplete ones (i.e., the greatest number of edges). Whenever a weekly purchase of shares is made, the person receives virtual rods for building a three-dimensional incomplete open cube. In the first week, the person is given three rods to form the first three-edge open cube. In the second week, the person acquires another three rods to try different arrangements, and they acquire a different three-edge open cube. In the third week, the person is informed that there is yet one more different three-edge open cube in the list. This time around they spend slightly more time, and finally a catalog of all possible three-edge open cubes is completed. This concludes the first episode. The person is aware that the next episode with various four-edge open cubes is waiting for them to explore. They then keep investing on time to collect four extra rods and begin to catalog the first four-edge open cube. Similar processes repeat every week, and variations of four-edge open cubes are found one by one. After five weeks, the person exhausts all variations and concludes the second episode. According to LeWitt's catalog resulted from the serial project, there are totally 122 incomplete open cubes, equivalent to two years and six months of weekly investment, in terms of both time and money.

Imported Relations for Blending

Every week, the person makes an electronic transaction to buy shares and receives virtual rods. They then make an effort to find another version of incomplete open cube that is not yet cataloged on the list. Once a new variation is identified, it is appended to the list. Different people would very likely uncover the variations in different order, and hence their catalogs would not be in the same order. This is reminiscent of some characteristics of a blockchain. Simply put, a blockchain is a distributed, decentralized digital ledger that facilitates the process of recording transactions and tracking assets. It is distributed because everyone on the same network can access or have a copy of the updated ledger. As a transaction occurs, it is recorded in a "block" of data, which is later linked to the previous block, thus forming a chain as the ledger. Without a central authority on the network to maintain consistency and integrity among all the ledger copies, a proof is needed for identifying and validating the next authentic block (containing authentic transaction records) that extends the chain. One very clever protocol that emerged is proof-of-work, used in the cryptocurrency bitcoin. It relies on a cryptographic hash function, which takes in any message data and outputs a fixed-length, seemingly random, sequence of bits, say 256 bits in case of SHA256(). It is infeasible to predict the output of an input, or reversely determine the input for a particular output, without resorting to wild guess and check. Consider a new block ready to extend the chain. To "find" a special number that, if added to the block, would lead to a hash output with certain pattern, for example, starting with 60 zeros in the 256 bits, requires a lot of computational effort. Yet, this special number provides good proof-of-work for validating this next block to the chain, because the required effort discourages fraud. The computationally demanding work, in case of bitcoin, is done by so-called "miners," who all compete to find a number for the next block and the winner gets a reward in return.[29]

Now reconsider incomplete open cubes. Finding a variation that is not yet cataloged by the person who practices the investment routine is comparable to the computational work of finding a special number input for the hash output requirement. The person might invoke the external knowledge frame [blockchain technology], which includes a relation FIND (solution, proof). The relation applies to both the computational process of finding a special number for the hash output and the person's cognitive act of finding a different variation of incomplete open cubes. When the person commits a new purchase of shares online and is given virtual rods for building new variations, one can imagine a third story of working to find a solution as proof for the new transaction just made. (Here, the solution can be hinted by other sources, because this "work" is kind of symbolic.) Adding a new variation to the list of incomplete open cubes represents appending a new authentic investment record to one's distributed ledger. Different people would likely find the variations in different order, resulting in different catalogs. Each person's catalog of incomplete open cubes is one's own symbolic blockchain of the investment routine. Table 6.7 summarizes the above blending processes.

Summing Up

People practices different routines for different reasons, including targeting a specific standard, fulfilling a checklist, or supporting a lifestyle grounded in a belief,

TABLE 6.7 The formal representation of the blends between the routine of dollar cost averaging and the narrative of *Incomplete Open Cubes*

Dollar cost averaging	Selective projection & pattern completion	Cataloguing variations of incomplete open cubes
PURCHASE (person, stocks) TRANSACTION ($100, shares)	→ PURCHASE (person, stocks) → TRANSACTION ($100, shares) [blockchain technology]: FIND (person, 3-edge open cube) LOG (person, 3-edge open cube) PoW (3-edge open cube) APPEND TO LEDGER (TRANSACTION ($100, shares), PoW (3-edge open cube))	REARRANGE (Sol, 3 edges) ← FIND (Sol, 3-edge open cube) ← LOG (Sol, 3-edge open cube)
...	...	...

which match different kinds of analogous stories. If a person is concerned with progress, a chain tale is appropriate. If one prefers variety and novelty, a theme-based narrative should be considered. In most cases, routines are supposed to run for an extended period, which entails more cycles than the iterative events of an analogous story. Hence, iteration at a higher level is suggested, that is, iteration of the analogous story into sequels, leading to a series. Iteration of a chain tale generates an episodic series, while iteration of a theme-based narrative forms an anthology series.

Each chain tale of an episodic series features a sequence of growing or piled-up events, which can be mapped with one's progress made in the same goal-directed routine. Between one tale and the next, the person practicing the routine might feel satisfied with the tentative conclusion and choose to pause. To keep the person motivated and prevent procrastination, building continuity between tales is the solution. This can be achieved by weaving a continuing storyline across all tales in the episodic series, that is, the intradiegetic approach. As informed by cases in film or television that combine episodic series with characteristics of serial narratives, a continuing quest or an enduring monster can be considered as an overall story arc. Apart from creative works in popular culture, serial art projects, typically involving a rule-based cataloging process that generates variations along a continuing progression, can also be interpreted as an analogous story for mapping and blending with belief-driven long-running routines, such as money saving or long-term investment.

On the other hand, each theme-based narrative of an anthology series explores variety under a consistent theme, which adds color to maintenance routines.

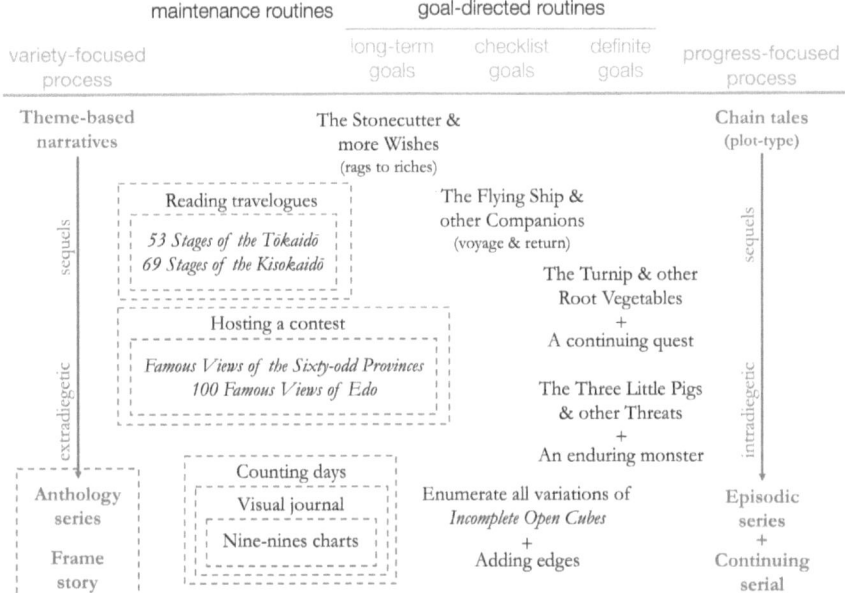

FIGURE 6.4   Distribution of series of analogous stories along the continuum between variety-focused and progress-focused routines.

Yet, the continuity between successive stories can be weak, because they only share an overall theme but each involves completely different characters and settings. Weaving an overarching storyline through them is not always feasible due to the major discrepancies between them. Enveloping them in a frame story, that is, the extradiegetic approach, is more practical. After finishing one theme-based story, the person is transported back to the frame story, whose unfinished storyline leads them to the next theme-based story. The classical examples in literature provide useful frame story ideas that maintain unfinished business on the audience's mind, such as an unfinished commitment or obligation. The frame story, as a metaphor borrowed from the visual art domain, can also be observed in the intriguing traditional Chinese calendar charts used for counting winter days. The chart contains two-dimensional visual and narrative frames on multiple levels, from writing each stroke of a pictographic or ideographic character as a sketch, to writing each character of a poem as a caption, to counting the days and documenting the weather throughout the season. This can be interpreted as a series of analogous stories enveloped in multiple frames, which match maintenance routines like walking one's dog every day. Figure 6.4 summarizes the ideas of episodic and anthology series of analogous stories, extradiegetic and intradiegetic approaches, and the case studies in this chapter along the continuum between variety-focused and progress-focused routines.

7
ITERATION OF VARIANTS

Routine involves iteration of actions. The iterative processes can be unexciting. Blending a routine with an analogous story turns it into expressive iteration. That is, through practicing the routine, a person tells an imaginative story to oneself or others. For different routines, one has to look for different analogous stories. Chapter 5 introduces different sources of analogous stories for mapping and blending with different kinds of routines. Chain tales of different plot types can be considered for goal-directed routines with emphases on a range of nuanced goals, from a specific standard on a scale, a checklist, to a long-term vision. Maintenance routines, which are usually driven by a belief instead of a goal, demand theme-based stories like travel narratives that highlight varied outcomes rather than repetitive actions. Chain tales foreground progress, while travel narratives feature variety. Meanwhile, analogous stories often have fewer cycles of iteration than most routines do. Hence, Chapter 6 suggests iterating analogous stories into different series for extended mapping with long-running routines. For an episodic series of chain tales foregrounding progress, the intradiegetic approach is used to enhance continuity across the series. For an anthology series of theme-based narratives, the extradiegetic approach is used instead. The framework, up to this point, seems comprehensive enough.

Yet, life can be unpredictable. On occasions, people might skip, suspend, or even quit a routine for various reasons, such as heavy workload from the daytime job, changes in living environments, health conditions, disruptions from other habits, and others. When a person stops engaging with a routine, one might not feel attached to the analogous story. After a period of time, one might have even forgotten about it. In case the person intends to resume the routine sooner, they might prefer seeing completely different analogous story simply for a sense of novelty. At times, an individual just feels disinterested in the original

DOI: 10.4324/9781003391449-9

analogous story, probably because of the genre, the topic, the characters, or the settings. For instance, one might find fables (i.e., stories involving animals as main characters) like "The Three Little Pigs" childish. Urbanites might find tales about farming (like "The Turnip") or stonecutting (like "The Stonecutter") difficult to relate to. There are cases where people find it difficult to be fully absorbed by the narrative and transport themselves into the story world. For instance, some may never identify with the protagonist of "The Flying Ship" who is too simpleminded. In case of travel narratives, someone might not be interested in the historical routes circa the Edo period. All in all, we not only need to look for appropriate analogous stories for turning a routine into expressive iteration but also want to identify a story that matches the user's demographics such as background, preference, and flavor. In reviewing the sources of analogous stories, fortunately, chain tales as a category of folktales, typically come with variants, which are varied versions of the same story. Variants deviating the least from the original story ride on a similar storyline and only differ in the ending or minor details. Other variants can be regarded as parallels or updates, which differ in their characters, settings, or even the story world, but they feature very similar plots, or specifically chains of repeating actions in case of chain tales. For theme-based stories such as travel narratives, variants can be slightly more ambiguous. Yet, iconic travel routes like the main roads connecting Edo and Kyoto actually reappear across various media forms, from prints, novels, to even board games. They are like counterparts in different domains. These variants, parallels, counterparts, or updates of analogous stories, which I collectively call "variants" for easy reference, offer a wide range of options for a person to choose from. With such expansive pool to choose from, one is likely to find one that matches their flavor. In this chapter, we will look at variants of analogous stories and how their genres, subject matters, characters, and even forms can be varied.

Apart from selecting a preferred analogous story from a pool of variants, an individual can also personalize a version of the story, based on knowledge acquired from the mapping between varied analogous stories and the routine. The vital relations in the analogics inform the most essential patterns that should remain in the new personalized version, while other components are replaceable. Consider the narratives of the traditional Chinese winter calendar charts discussed in Chapter 6, which feature different poems. Every poem can be regarded as a variant of the analogous story for mapping with a daily routine. Common to all variants, all poems are composed of nine characters, and all characters are made of nine strokes, with each stroke corresponding to each day. These are essential patterns or rules, what I call the "nexus" elements, defining this group of variants. The theme, content, representation, and tone of a variant can be personalized. In this chapter, I will discuss and demonstrate how we can update or personalize a version based on an analogous story, which can be a chain tale, a travel narrative, a calendar chart, or even a catalog based on Sol LeWitt's serial art project.

Variants of Chain Tales

Folklorists call different versions of one folktale "variants."[1] As mentioned in Chapter 5, folktales are stories orally passed down along generations in a culture. When a story has been told again and again, it inevitably changed. When it was transcribed by different people at different times in different contexts, it spun off different versions. The transcription process could also entail modification and involve creativity. For instance, the Grimm Brothers, as philologists, seemed to more faithfully record their collection including raw and gritty descriptions of the harsh reality faced by people of their time, while Joseph Jacobs tended to have purposely modified the stories for the benefits and taste of his primary target audience, children. Andrew Lang's tales are usually more elaborate in details and images.[2] Sometimes, a folktale might spread widely across different cultures. Hence, it is commonplace to see strikingly similar storylines or other recognizable patterns among folktales from different regions. Folktales with numerous variants across the world include "Cinderella," "Little Red Riding Hood," "The Three Little Pigs," and many others. The number of variants, say for "Cinderella," can reach as many as 900.[3] The many variants of folktales provide valuable resources for folklorists' comparative studies. In this chapter, I do not intend to perform or repeat a comprehensive comparison of variants like what folklorists typically do. Instead, I focus my analyses here on those variants that hint at how we can update or personalize an analogous story to better match one's preference or temperament.

Some variants might only differ slightly in the ending of some characters. For example, in Jacobs's version of "The Three Little Pigs," the first and second pigs are unfortunately eaten by the wolf and the third pig finally boils the wolf and eats it. In Lang's version in *The Green Fairy Book,* the fox, the variant villain, catches the first and second pigs and locks them up in its den. The third pig finally saves its siblings after killing the fox. Today, the storyline is simplified in many versions so that the first two pigs just run to the third pig's house after their own houses have been blown down by the wolf. Some variants of one folktale might have completely different characters and settings, and the remaining recognizable similarities are the chain of repetitive events. As mentioned in Chapter 5, "The Fisherman and His Wife" in the Grimm Brothers' collection and "The Stonecutter" belong to the same Aarne-Thompson-Uther (ATU) type 555. Both tales are composed of a chain of repetitive and increasingly ambitious requests, but their characters and settings are completely different. They also have slightly different endings in terms of the protagonist's awareness of values. Another prominent example is a group of tales classified as ATU type 513A, including "How Six Made Their Way in the World" and "The Six Servants," both collected by the Grimm Brothers, together with "The Flying Ship." Their respective protagonists have different backgrounds and archetypes, but they all set off on a voyage and meet a number of talented companions, which would help them to earn rewards and return home. These variations, rather than variants, may be called parallels, as they are like similar versions in parallel story worlds.

In Chapter 5, several classical chain tales featuring particular plot types are discussed in relation to their roles as analogous stories for mapping with different kinds of routines. In this section, variants or parallels of these chain tales are selected for showing the possibility of transforming the character or the setting of an analogous story, while maintaining the blend with a routine.

Variations from Animals to Humans

In considering a story for mapping with a routine, some people might find fables, typically with animals as main characters, childish or even silly. "The Three Little Pigs," among the chain tales discussed in Chapter 5, is a well-known fable. Although many variants have emerged, the pigs and the wolf, or related animal species, remain the main characters. Meanwhile, it is not impossible to replace the pigs with human characters, because their behaviors in the fable, such as building houses and boiling water in a pot, are already anthropomorphized. Among many modern variants, *The Three Little Pigs: An Architectural Tale* by Steven Guarnaccia (2010) successfully transforms the three pigs into three architects (though still illustrated with a piggy head), who build their homes alluding to three real-world celebrated architects' signature houses, namely Frank Gehry, Phillip Johnson, and Frank Lloyd Wright. Gehry has told the media about his inspiration from childhood when he built little cities out of scraps of wood and other materials that his grandmother brought him from his grandfather's hardware store.[4] The first fictional architect in Guarnaccia's book thus builds a house of scraps. Johnson's signature work is the Glass House, a one-floor house in a minimalist steel frame with all exterior walls made of glass. The second fictional architect thus builds a house of glass. Finally, one of Wright's famous works, Fallingwater, is a house built over a waterfall, and the construction materials include stone and concrete, which are the materials used by the third fictional architect. Guarnaccia's version intelligently updates the materials for the three houses in the story. That said, the villain is still a wolf, though in a gangster costume. For a complete transformation away from the fable genre, I suggest replacing the animal villain with natural disasters, such as the first house of scraps being blown down by a storm and the second house of glass being destructed by a tornado. The three fictional architects finally stay in the third house until the sky clears up again. The plot can still be counted as "overcoming the monster," with the monster coming from the extreme weather. In summary, the new version maintains the same plot, similar repetitive events (i.e., construction and destruction), which are the nexus elements, but changes characters from fictional animals to real-world architects as well as changes some objects in the story (i.e., building materials).

As discussed in Chapter 5, "The Three Little Pigs" is a chain of iterative construction works using increasingly more resilient materials and facing increasingly stronger attacks, which remains as is in the version involving architects instead of pigs. The protagonists' goal is the same, that is, to build a house strong enough to stand the attacks. The updated version can also be mapped and blended with a goal-directed routine like taking medications or supplements to boost immune system, or better still, parents getting kids to take medicine. The external knowledge frame brought in for the blend, as suggested in Chapter 5, [neighborhood] still applies. The imported relation NEAR (home, new house) still works in binding together a person's real home and each fictional architect's house. When the person takes a pill, they virtually see the fictional architect building a new house near their own home, maybe through a wall-mounted digital display simulating a window. After taking the pill several times, the architect's house is completed. Yet, a virtual storm then blows it down. Another imported relation CAUSE (BLOWN (), SICK ()) reminds the person that the gusts nearby would cause one to catch a cold. The person can also imagine a third story of strengthening the body and helping the architect's neighbors to build a resilient house against the blows. The new non-fable version can be considered an analogous story for those who prefer human characters in a "modern" context.

Variations on Character Archetypes

Most chain tales discussed in Chapter 5 engage human characters who are archetypes, such as a prince in "Tsarevitch Ivan, the Firebird and the Gray Wolf" or a simpleminded third son in "The Flying Ship." In the blend of a chain tale with a routine, a person imagines a third story and takes the protagonist's perspective, which leads to strong persuasive effects. Yet, some people might find it difficult to put themselves in the shoes of a particular character type, such as a simpleminded person. There are other chain tales whose settings and plots are similar to those of "The Flying Ship" but the protagonists are of very different types. In "The Six Servants," a German folktale collected by the Grimm Brothers, the story shares a similar setting with "The Flying Ship," that is, an open contest for marrying the princess. But in "The Six Servants," it is the prince who intends to take the challenge. In another tale also collected by the Grimms, "How Six Made Their Way in the World," a soldier not satisfied with the scanty compensation he receives after military service, wants to inherit the king's wealth. He joins the competition in order to marry the princess. All three protagonists, respectively, meet and build their teams of unique talents and peculiar powers. Although each team has slightly different talents and faces slightly different challenges in their respective stories, they all go through similar events where each team member solves at least one problem and the full team safely return home in the end. In short, the three tales are parallel in their chains of recurring events (i.e., recruitments of different talents and tackling different problems), the setting (i.e., a contest for marrying the princess in a kingdom), and the plot type "voyage and return" (i.e., winning the contest), which make up the nexus. Yet, they varied in the character type with the simpleminded son, the prince, and the former soldier being the protagonist.

"The Flying Ship" is suggested in Chapter 5 to map with routines performed for checklist goals, such as attending regular classes for a period of time. When a person attends one session, they virtually recruit one talented member into their team. The person can invoke external knowledge frame [team project] and import new relations including PEER () and LEARN FROM (), imagining a third story of working with talented classmates with different skills for the final project. If the person cannot identify with the simpleton son in the story, they might consider being the prince in "The Six Servants" or the former soldier in "How Six Made Their Way in the World." Even though a different tale is chosen as the analogous story, one can still use the same external frame and imported relations in the blend with the routine. Whenever acquiring a new skill in a class, the person can imagine recruiting a new member with unique talent. There is no flying vehicle for the prince or the former soldier, but that is not necessary for the blend to take place.

Variations on the Setting

Some variants can be so different that not only do the characters change but also the whole setting. "The Stonecutter," the Japan-originated tale discussed in Chapter 5, has a distant yet comparable parallel in Germany, "The Fisherman and His Wife," as collected by the Grimm Brothers. The protagonist in the former meets a mountain spirit when cutting stones in a quarry. In the latter, the protagonist catches a fish who grants him wishes. The stonecutter repeatedly asks for transformation into more powerful beings. The fisherman, driven by his wife, makes increasingly greedier wishes for wealth and power. Both the stonecutter and the fisherman (and his wife) return to their original selves at last, after repeatedly making bigger and bigger wishes. Although the two tales differ entirely in their settings (in the mountain vs. by the sea) and the characters' occupations (stonecutting vs. fishing), their largely similar chains of events make them belong to the same plot type "rags to riches," which is the nexus. A subtle difference lies between the two endings. The stonecutter finally is enlightened and becomes aware of his own value in the circle of cause and effect, while the fisherman and his wife remain in despair.

For mapping with routines driven by a vision or belief like trading used items for new goods or recycling materials, both tales can be an analogous story. As suggested in Chapter 5, the external knowledge frame for the blend is [chain reaction], which includes new relation PROPAGATE () that distantly connects one's act of trading used items for new goods with the other's act of exchange for new power or new possession. If a person who regularly trades used items finds stonecutting in a quarry too distant from their true selves, they could consider if fishing by the sea is a better option. With the same external frame invoked and new relation imported, the person can use the fisherman tale, or update it to someone fishing by the river in a city, as the analogous story for the blend and imagine a third story. In the blended story, the person's act of trading used items triggers a chain reaction leading to the characters' transformation to grab more power.

TABLE 7.1 Sample variants and their variations on different components

Variant	Plot type	Main characters	Other existents	Setting
The Three Little Pigs	Overcoming the monster	Pigs	Building materials: straw, wood, bricks	Building a home
The Three Little Pigs: An Architectural Tale	Overcoming the monster	Architects	Building materials: scraps, glass, stone and concrete	Building a home
The Flying Ship	Voyage and return	Simpleminded son	Seven companions	Traveling to compete for marrying the princess
The Six Servants	Voyage and return	Prince	Six servants	Traveling to compete for marrying the princess
How Six Made Their Way in the World	Voyage and return	Former soldier	Five servants	Traveling to compete for marrying the princess
The Stonecutter	Rags to riches	Stonecutter	Wishes: a rich man, a prince, …	Stonecutting in a quarry
The Fisherman and His Wife	Rags to riches	Fisherman	Wishes: a nice house, a king, …	Fishing by the sea

Insights from Variants of Chain Tales

The above variants demonstrate possibilities of iterating a chain tale with variations on different aspects while blending with a routine. Table 7.1 summarizes the observations. Some of them have the main characters varied between animals and humans, or from one type to another, as well as slight differences in minor characters or objects, collectively called "existent" in Chatman's terms (1978). Some even have different settings. However, these variants or parallels have similar chains of events that can be mapped and blended with the same routine. Most importantly, the same external knowledge frame can be invoked for the blend. The consistency in framing allows a person to perform the blend for imagining the third story, when one chooses to change the analogous story for personal preference.

These sample variants also cast light on updating or personalizing a chain tale for mapping with a routine. Their respective varied aspects show ways of changing the characters, objects, or even the whole setting that might respond to individuals' preferences, as long as the changed components remain compatible with the invoked knowledge frame. I summarize the insights in the following three strategies.

Looking for Real-World References

The Three Little Pigs: An Architectural Tale creatively modifies the main characters, the three little pigs, to be more human-like. They walk on two feet, wear styled outfits, and design their own houses, while keeping their piggy faces likely to appeal to younger target audiences. Their outfits and choices of materials actually refer to three well-known architects in the real world who use signature building materials in their respective works. This fits perfectly in the original chain of repetitive acts of construction using increasingly resilient materials. The new three materials, namely, scraps, glass, as well as stone and concrete can substitute the old ones, namely, straws, sticks, and bricks. This modern variant keeps all other parts of the story intact, including the repeated constructions and destructions, and hence the invoked knowledge frame [neighborhood] and imported relation NEAR () definitely remain valid for binding together the medication-taking routine and the fictional architects' houses. Even with my suggestion of replacing the wolf villain by natural disasters such as a storm or a tornado, they still apply. The insight from this modern variant is that an individual who has to take medicine or supplements regularly can update or personalize the classical chain tale by referring to people, artifacts, locations, or facts that one knows pretty well in the real world.

Considering Different Social Groups

In the variants "The Flying Ship" and "The Six Servants," the protagonists are folklore archetypes. A simpleton or a prince recurs as the main character in many folktales. Meanwhile, "How Six Made Their Way in the World" seems to employ a slightly alternative character type, a soldier discharged from military service. Unlike a simpleton or a prince, which sounds distant to a reader of today, a former soldier is more relevant in contemporary popular culture. The character type, for instance, evokes of the hero warrior retired from the Vietnam War featured in the *Rambo* film series (1982–2019). In fact, it represents a niche social group. In social identity theory, a social identity is defined as "a person's knowledge that he or she belongs to a social category or group" in which a social group is comprised of "a set of individuals who hold a common social identification or view themselves as members of the same social category."[5] In this sense, a veteran, who might no longer be regarded as "soldier" but has not yet developed a sense of belonging in any other social category, would be in a special status of "former soldier." Similar social categorization might extend to someone who has retired from a profession, say "soccer player," but has not yet started another new career, and they might actually consider themselves a "former soccer player." In "How Six Made Their Way in the World," the former soldier recruits a team of talented members and continues his voyage for an honor (i.e., to win the contest and marry the princess). In the present day, a former soccer player might recruit talented players to form a new team, like David Beckham's Inter Miami CF recruiting Lionel Messi in 2023. The insight here is that one can update or personalize a chain tale of iterative recruitments by giving the protagonist a special former identity that drives them to recruit a team with different talents in pursuit of an unfulfilled desire. This mission should of course match the original blend with a routine toward a checklist goal. A former team player of any kind perfectly fits the invoked knowledge frame [team project].

Identifying a Profession

As the title says, the protagonists in "The Stonecutter" and "The Fisherman and His Wife" are, respectively, a stonecutter and a fisherman. The stonecutter in the end realizes the unique power of stonecutting: the ability to cut a massive mountain into rocks. The fisherman, on the other hand, applies his skills to catch a magical, omnipotent fish. Both protagonists are in fact skillful in stonecutting and fishing, respectively. In the real world of today, these jobs are supported by technology and thus require highly specialized technical skills, although some people might not consider either of them a "professional" nowadays. Yet, a stonecutter or fisherman by profession can make a successful career, as suggested by various stonecutters in long-term partnership with the Renaissance master Michelangelo[6] and the successful shrimping business founded by Forrest Gump, though the latter is fictional. More importantly, a talented practitioner in such modern skill-centric professions, such as a chef or hairdresser, is much sought after and is able to climb the hierarchal ladder quickly. The escalation in the professional status is intriguingly comparable to the fictional stonecutter's or fisherman's increasingly ambitious requests for power. Where is the end of the ladder? When will one consider oneself successful within a trade? These questions are worth contemplating. In fact, a person practicing routines matching chain tales of "rags to riches" can consider a preferred profession other than stonecutting or fishing for the analogous story. As long as the job leads to ascent in social status, the originally invoked knowledge frame [chain reaction] and imported relation PROPAGATE () still apply.

New Variants of Chain Tales

Based on the above three strategies, I suggest the following modern variants as a demonstration of how one might update or personalize an analogous story for blending with a routine. Each of the following variants basically incorporates ideas inspired by some of or all three strategies.

How Eleven Made Their Way in the League

If a person practicing a routine with a checklist goal, such as attending a community college course, finds it hard to relate the traditional chain tales of "The Flying Ship," "The Six Servants," or "How Six Made Their Way in the World" to their daily life, they might consider the three aforementioned strategies together with the invoked knowledge frame [team project]. First, identify a profession that involves building a team, such as professional sportspersons. Second, identify a special identity group that exists within that profession including, say, someone retired. Third, look for real-world references. An example can be soccer players who are retired and then turned themselves into successful coaches or team managers, such as Johan Cruyff, Pep Guardiola, Zinedine Zidane, and many others. Guardiola, for instance, was a great midfielder during his playing days, winning six La Liga titles in addition to an Olympic gold medal with Spain in 1992. As a manager, he won multiple league titles with Barcelona, Bayern Munich, and Manchester City. A personalized and updated variant can be titled "How Eleven Made Their Way in the League." It is about a footballer-turned manager who recruits talented players for different positions (e.g., goalkeeper, backs, wings, midfields, center-forward, striker) in a soccer team and aims at winning the championship in the community league. When the person attends a class in real life, they meet a virtual classmate who is a former footballer now recruiting team players. After each class, this class-mate would have successfully got one position filled by someone. Toward the end of the course, all 11 positions are filled and the full team is heading to the community league. This story sounds familiar, doesn't it? Yes, similar variants can be found in manga series like *Slam Dunk* (1990–1996).

The Chef

Now imagine a person engaged in a routine of trading used items like clothes or handbags for new items, which is driven by a vision of sustainability. The traditional chain tales "The Stonecutter" and "The Fisherman and His Wife" might not easily be linked to the urban lifestyle of trading used items, even though the person invokes the external knowledge frame [chain reaction] and tries to imagine the connection. In fact, stonecutting and fishing might not be common professions in big cities. One might look for another urban profession that entails a hierarchal ladder for its practitioners to climb by changing jobs. A chef is an urban profession, and getting employed by another restaurant is a typical way to accumulate experience and climb the hierarchal ladder in the field. A prominent and recent example is André Chiang hailed from Taiwan. His story is turned into a documentary streamed on Netflix.[7] Chiang started his relationship with the kitchen when he was a kid helping out in his mother's little restaurant in Japan. As an adult, he moved to France and spent nine years in a Michelin three-star restaurant to become the chef. He then worked in different restaurants in France, Thailand, China, and Seychelles, before opening his own venture André in Singapore in 2010, which won him many awards including the honor of owning two Michelin stars. While all people around him anticipated his next move was to earn "one more Michelin star," he closed the restaurant in 2018 and focused on another restaurant of his, named "RAW," in his hometown. He even returned his Michelin stars and requested that RAW not being included in the Taiwan Michelin Guide. Chiang's culinary career can be a real-world reference to an updated variant used for mapping with one's trading routine. Whenever the person trades a used item for a new good, one might imagine that the act of trading triggers a chain of exchanges, probably including someone bringing a designer handbag to a fine-dining establishment, which leads to a chef quitting to join another Michelin-starred restaurant. At certain points, the chef suddenly gets their priorities straight like Chiang or the stonecutter, recalling why they joined the profession in the first place. The person trading their used items may also rethink their values regarding possessions and necessities.

The Fallen Tree

The Russian folktale "The Turnip" has been retold or translated in many other cultures, but the variants are largely similar. If a person who performs a routine toward a specific goal, like regular workouts to achieve an optimal body weight, is not interested in a tale of farming vegetables, one might give thought to other kinds of manual work that also involves increasingly more people. Jobs like construction or renovation require hard work but commercial projects typically engage a fixed number of staff as budgeted. Instead of commissioned jobs, volunteer work might have more flexibility. A special type of volunteer work requiring substantial physical strength is hiking trail maintenance.[8] Trail maintenance includes a variety of hard work, such as operating chain saws to cut fallen trees, rolling the cut logs off the trail, removing branches, and clearing overgrown vegetation on the trail. Related service groups always call for volunteers. Experienced volunteers with physical strength can operate chainsaws or roll the cut logs, while new or young volunteers can also help remove branches or pull out overgrown vegetation. Thus, a variant about this special volunteer group can be developed, titled "The Fallen Tree." When the person finishes their daily workout, they virtually see a stocky volunteer cutting a large fallen tree that blocks a hiking trail. On another day, the person sees one more volunteer joining to roll the cut logs off the trail. On the following days, they will see volunteers of different ages and sizes joining the crew to remove branches and pull out overgrown vegetation. More volunteers can be added to the chain, until the trail is fully restored and open again. The originally invoked knowledge frame for the blend [urban farming] no longer applies there. Instead, there are trails in urban areas, and the person may imagine a trail near the gym where they work out. Hence, the person can see the fallen tree blocking the trail. The external knowledge frame here can be [city trails].

Variants of Travel Narratives

A chain tale is composed of a chain of causally linked repetitive events. The actions (e.g., pulling up the giant turnip) are sequentially presented with "gradual growth" in intensity (e.g., using more durable materials) or "piling up" of items (e.g., more people getting on board), which feature progress and render the tale suitable for mapping with goal-directed routines. For maintenance routines where variety, not progress, is preferred, a narrative whose recurring events are only implicated between scenes can be considered, because it highlights varied intermediate stages (e.g., arriving at a different station) rather than repetitive actions (i.e., walking on the road). In Chapters 5 and 6, I introduce examples of theme-based scene-to-scene narratives. Many ukiyo-e woodblock print series, such as *53 Stages of the Tōkaidō*, can be regarded as scene-to-scene travel narratives. Sometimes, a well-received print series in the Edo period was followed by sequels, constituting an anthology series of print series, in other words, a thematic collection of scene-to-scene travel narratives. For instance, *53 Stages of the Tōkaidō* is a travel narrative on the theme of the Tōkaidō, while its sequel *69 Stages of the Kisokaidō* has its theme on another road, the Kisokaidō (aka the Nakasendō). The two travel narratives share a larger common theme, which is traveling between Edo and Kyoto on a road. When a person performing maintenance routines like doing the laundry at home has finished the travel narrative on the Tōkaidō, they can continue with another narrative on the Kisokaidō. Both narratives are within the larger frame story of reading travelogues. In case the person becomes bored of travelogues, is there another variant for them to choose from? Unlike folktales, "variant" is not a common term used among published travel narratives in prints. Yet, different versions of Edo-period travel narratives, say on the Tōkaidō, can be found in some other forms, not in printed postcards but printed board games. In this sense, I still regard these different versions as variants. They all share the nexus, that is, the 53 stations on the Tōkaidō.

Dōchū Sugoroku as Travel Narratives

With the advent of woodblock printing in the Edo period of Japan, a kind of board game, called "sugoroku," or "e-sugoroku" to be specific, became popular. Sugoroku (双六), with two kanji characters meaning "double six," can refer to two different forms of board games, including one played on a wooden board similar to backgammon, called "ban-sugoroku" (board sugoroku) and the other played on a printed sheet of paper, called "e-sugoroku" (picture sugoroku). The discussion here only covers the printed picture form and I refer to it as simply "sugoroku" for easy reference. Sugoroku is a competitive race game similar to snakes and ladders. Players throw a dice in turn and move their tokens accordingly on the printed paper board. The printed paper board includes an elaborate, typically spiral, path for players to race from the starting point to the finishing point. The origin of sugoroku is unclear.[9] The earliest sugoroku in Japan possibly dated back to the thirteenth century printed in black ink on paper with Buddhist teachings and it was supposed to be a teaching aid for novice monks. During the Edo period, the advanced woodblock printing technology enabled more colorfully printed and visually elaborated sugoroku designs. The themes of the game also became varied with Buddhist teachings, Confucius wisdom, social codes, career ladders, and theatrical entertainment being the subject matters. A printed path of sugoroku was

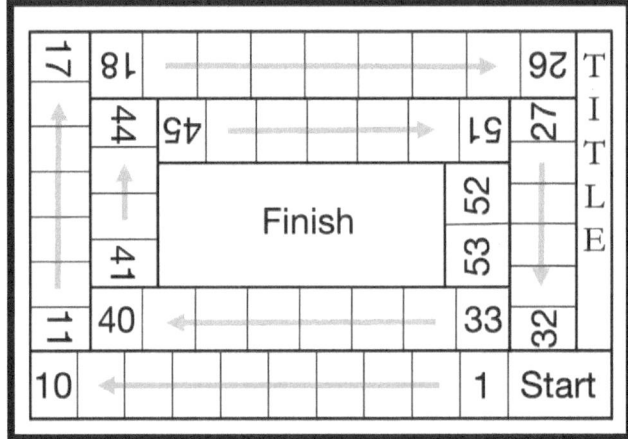

FIGURE 7.1   A wireframe (made by the author) of a typical dōchū sugoroku featuring a rectangular spiral path.

often a sequence of squares representing a metaphorical journey of one's life with each individual square being a different stage in life. As ukiyo-e single sheet prints appeared after the mid-Edo period, a type of sugoroku with a travel theme also emerged, called dōchū (道中 meaning "on the road") sugoroku. The game's path covers a travel route with a miniature road scene in each square. Among those prints of dōchū sugoroku surviving till this day,[10] the most common format is a rectangular spiral path starting from the bottom right corner, and ending in a finishing point at the center (see Figure 7.1). The most popular travel route featured is the Tōkaidō, starting from Nihonbashi in Edo, through 53 stations, to Sanjō Ōhashi in Kyoto, represented by totally 55 squares on the paper board. Imagine a player moving the token square by square according to the thrown dice. As the player's gaze sweeps across the miniature street scenes in each square, they feel like virtually visiting different stations on the road. Due to the randomness in dicing and the occasional need to skip a turn in the square marked with a kanji meaning "overnight stay" (泊), players tend to traverse the same path at different paces. Comparing with browsing through the 55 ukiyo-e single-sheet prints of the series *53 Stages of the Tōkaidō* in a prescribed order, a sugoroku player's journey across the 55 miniature street scenes is like a different version of a similar travel story unfolding in a different medium. As there are many different published designs of dōchū sugoroku on the theme of the Tōkaidō, all of them and the Hiroshige's ukiyo-e print series can be regarded as variants. Among the many designs, several of them are particularly distinct variants whose aesthetics offer players different experiences of traversing the road, and thus they can be considered differently for mapping with the same routine.

Although the artistry of sugoroku boards has been less highly regarded than ukiyo-e prints,[11] ukiyo-e masters like Hiroshige also designed quite some sugoroku boards. A few of Hiroshige's sugoroku works intriguingly demonstrate variations of visual arrangements of the travel route on the paper, from the typical rectangular spiral of separate squares to a similar spiral where the squares show no borders, and a completely irregular spiral path embedded in a bird's-eye view map. Let us look at these variants in the following paragraphs.

*Tōkaidō Gojūsantsugi (53 Stages) Sugoroku (*東海道五十三次雙六*)*

This sugoroku game board was designed by Utagawa Hiroshige and published by Yamajin, but the year of publication is unknown. The dimensions are 25 centimeters in width and 36 centimeters in length. The design of the spiral path is a typical example of dōchū sugoroku. The spiral is rectangular, starting from the first square at the bottom right corner, which is the starting point of the Tōkaidō, going to the left along the bottom edge of the board, consisting of 11 separate squares each of which presents one road scene at a station with its name clearly written at the top of the square. This horizontal segment of 11 squares running from right to left at the bottom edge is like a reel of photo shots taken at each station (see Figure 7.2). After the first 11 squares, the path takes a 90-degree turn to run from the bottom to the top along the left edge of the board, presenting the views of another seven stations. The pictures shown in these seven squares are also turned 90 degrees clockwise, and hence the player is supposed to move to the left of the board and see the board

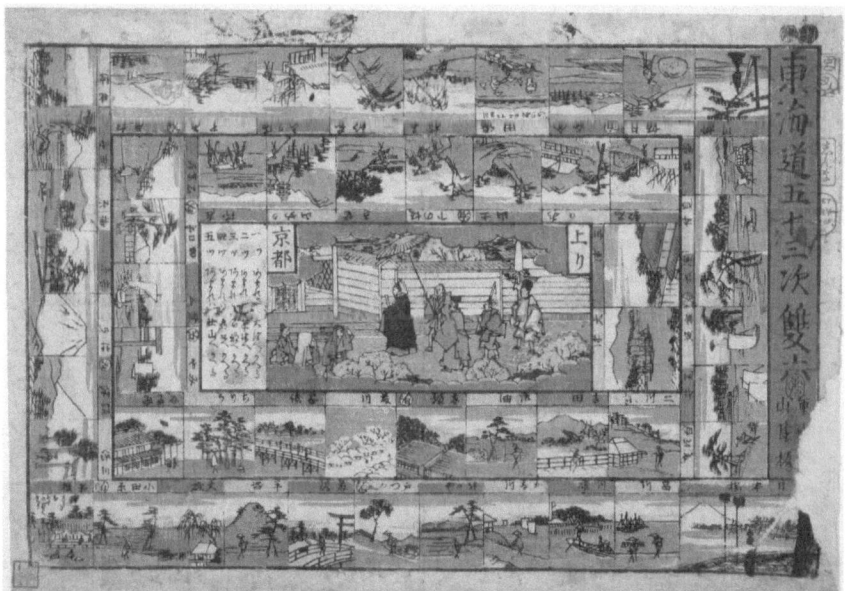

FIGURE 7.2   A digital archive of *Tōkaidō Gojūsantsugi Sugoroku*.

Source: East Asian Library, University of California, Berkeley.

from that side accordingly. The eleventh station, Mishima, sees only trees and a torii of a shrine. From the twelfth station, Numazu, to the fourteenth station, Yoshiwara, the image of Mount Fuji gradually enters the frame, appearing at the right, center, and then left of the picture, simulating a traveler's viewpoint moving from the left side to the right side of the mountain. These location-based views subtly echo what are presented in the ukiyo-e prints of the original Hiroshige's Tōkaidō series. In *53 Stages of the Tōkaidō*, the eleventh station, Mishima, also shows a torii in the mist, the thirteenth station, Hara, prominently features Mount Fuji in the center, and then the fourteenth station, Yoshiwara, includes the mountain in the background on right side. The sixteenth station, Yui, both in the board game and in the print series, shows a similar view with the mountain in the background from a distance. The spiral path then turns 90 degrees clockwise again and runs along the top edge of the board. Here, what seems like road-side restaurant is featured in the 20th station, Mariko, resembling the famous one at the same station depicted in the ukiyo-e print series. The spiral path continues to turn and runs for a shorter length after each turn, finally reaching the destination at the center. All in all, most miniature scenes in the board game resemble those shown in the single sheet ukiyo-e prints. The visual journey of a player of the former is comparable to that of a beholder of the latter. The two artifacts can be regarded as variants of the same travel narrative on the road.

In Chapter 5, I suggest that *53 Stages of the Tōkaidō* can be an analogous story for mapping with maintenance routines like doing the laundry at home. A person after finishing the routine can imagine a third story of folding and packing clean laundry in a guesthouse on the road and getting ready for another leg of the trip to the next station. Other than virtually seeing the ukiyo-e print of the next station, the person might instead virtually get a new stamp on the sugoroku board at the next station. In other words, the person can imagine carrying the paper board and collecting all the 55 station stamps on the road. The invoked external knowledge frame [accommodation] and imported relations still apply. Additionally, another knowledge frame [travel stamp] can be brought in with new relation STAMP (station, square) that binds together every station stamp and the corresponding square on the sugoroku board. This sugoroku is a stamp-collecting variant of the travel narrative.

*Tōkaidō 53 Stages New Board Dōchū Sugoroku (*東海道五拾三次新板 道中双六*)*

As the title suggests, this sugoroku board should be a new version designed by Hiroshige. The actual size almost doubles that of the previous one, measuring 42 centimeters in width and 62 centimeters in length. With a larger board, the printed pictures depict more details in the landscapes, trees, houses, and people, as well as the atmospheric textures in the sky. The spiral path also follows the typical rectangular format. Yet, the borders between consecutive squares disappear in this new version, with only a vertical station title bar partly separating two squares. The vertical title bar is about three-quarters the height of each square, leaving the

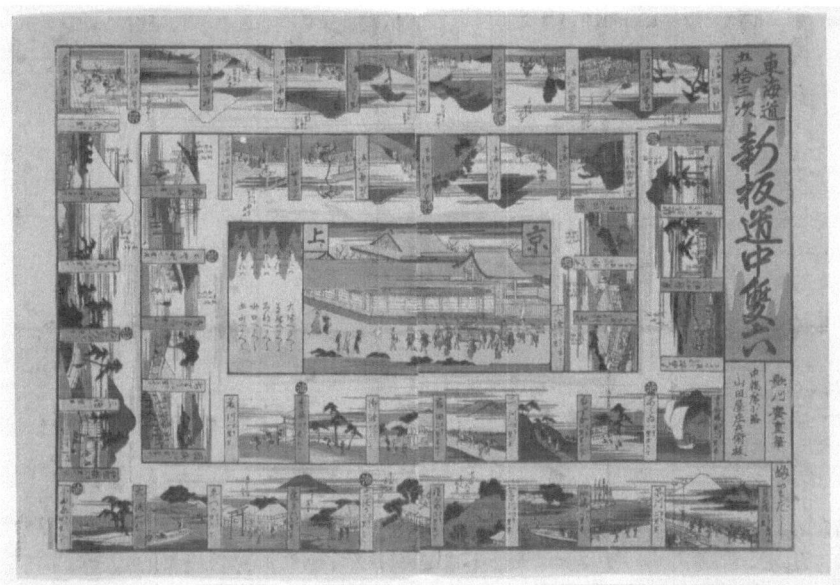

FIGURE 7.3   A digital archive of *Tōkaidō 53 Stages New Board Dōchū Sugoroku*.

Source: Tokyo Metropolitan Library, https://archive.library.metro.tokyo.lg.jp/da/detail?tilcod=000000 0004-00000060.

upper one-quarter of all squares connecting. Hence, visual contents like the sky, clouds, and some high mountains run continuously across adjacent squares (see Figure 7.3). For example, Mount Fuji stands in the background across the thirteenth station, Hara, and the fourteenth station, Yoshiwara. The separation of the roads, seashores, or houses in the foreground by the vertical title bars and the continuation of the mountains, clouds, and sky in the background create a unique visual aesthetic. It visually simulates the motion parallax phenomenon in perception, that is, from the perspective of a moving observer, objects that are closer appear to move faster than objects that are far away. An obvious example is viewing the landscape through a window on a moving train. The effect is already well simulated in animation. Consider a player of this sugoroku board game who moves the token as well as their eyesight across the squares on the path. Whenever the eyesight passes a vertical title bar, the foreground image "jumps" from one to another while the background seems to pan smoothly horizontally. It is like a speed-up version of motion parallax. Closer objects already move out of the frame, and others jump in. Distanced objects move relatively smoothly to the right. The player feels as if they are moving to the left and looking into the miniatures with an illusion of depth.

The spiral path of this new sugoroku board consists of 55 vertical title bars marking the 53 stations plus the starting and finishing points. In each linear segment, horizontal or vertical, the number of miniatures is fewer than the number of vertical title bars by one. The spiral path totally shows only 48 miniatures, with some station scenes skipped. That said, some remaining station scenes can still be recognized for their close resemblance with their counterparts in the ukiyo-e print series, such as the famous road restaurant at the 20th station, Mariko. Hence, this sugoroku board game is another variant of the same travel narrative analogous to the laundry routine. When a person finishes a laundry routine at home, they can imagine packing the clean laundry and moving on to the next station. Instead of seeing a ukiyo-e print or getting another station stamp, they might virtually move to the next square on the sugoroku path and see a different view. The discrete change in the foreground together with the continuous transition in the background simulate high-speed motion parallax and render a sense of horizontal movement. An external knowledge frame [motion parallax] can be brought in, together with the originally invoked external knowledge frame [accommodation]. They result in the blend, that is, carrying clean laundry and moving fast on the road to the next station. This sugoroku is a motion-parallax variant of the travel narrative.

Ukiyo Dōchū Hizakurige Comic Sugoroku (浮世道中膝栗毛滑稽双六)

This sugoroku board was designed by Ichiryūsai Hiroshige (一立斎広重) in 1855, said to be an alias of Utagawa Hiroshige.[12] Instead of depicting the landscapes along the Tōkaidō, it illustrates two main characters traveling on the road, based on a famous novel, *Tōkai Dōchū Hizakurige* (東海道中膝栗毛), written by Jippensha Ikku (1765–1831) and published in serial format from 1802 to 1814. The novel is a comical account of two young men named Yajirobei and Kitahachi who travel on the Tōkaidō from Edo to Kyoto on foot. In the title, the word "hiza" means "knee" and "kurige" means chestnut-colored horse, implying the characters use their own legs in place of a horse on the trip. Due to their "eccentric" characters, the two protagonists always run into a peculiar situation at every station, which is told accordingly in every chapter of the novel. This sugoroku, using the typical rectangular spiral format, shows in each square an illustration of the corresponding chapter of the novel. It measures 51 centimeters in width and 72 centimeters in length. The starting point, at the bottom right corner, is Kanda Hatchobori instead of Nihonbashi, because the two protagonists lived there, with the first illustration showing them discussing their plan of a journey, with someone outside eavesdropping. The first horizontal segment along the bottom edge of the board consists of seven illustrations (including the starting point), which look like, as opposed to a film reel, a comic strip.[13] Readers of the novel should be able to recall the fictional moments elicited by the illustrations, say, why the two protagonists are dragged along by two females in the street at the fourth station. Along the spiral path, many squares have illustrations that look amusing or hilarious. In the novel, the protagonists also visited Ise Grand Shrine, which is a branch from the Tōkaidō, before arriving at Kyoto. Yet, this sugoroku sticks to the original 53 stations of the Tōkaidō and skips some of the late chapters of the book. It can be regarded as an adaptation variant of the travel narrative shifting focus from landscapes onto two particular travelers.

This comic strip-like sugoroku can be an analogous story for the laundry routine. When a person folds the clean laundry, one can imagine leaving a guesthouse and heading to the next station, with the invoked knowledge frame [accommodation]. The person then virtually sees the next square on the sugoroku board. It is like a comic artist traveling on the same road coming across the two protagonists. The artist assumes an observer's perspective, that is, not involved in their "mess," but instead documents the moment in pictures. A new knowledge frame that can be brought in for the blend is [comic artist] and a new relation to be imported is WALK-PACE (artist, Yajirobei & Kitahachi), which means that the artist is traveling at the same pace as the two protagonists. Starting the journey at the same time with the comic artist, the person performing the routine can thus imagine themselves on the spot as the comic strip is drawn, while doing the laundry at home in reality.

Pilgrimage Up-Capital Dōchū Ichiran Sugoroku (参宮上京道中一覽雙六)

This sugoroku game board, designed by Hiroshige and published in 1857, is a huge one compared with others. It is about 72 centimeters in width and length. Instead of following the typical rectangular spiral format, the design intriguingly embeds an irregular path in a bird's-eye view map. The path still starts at the bottom right corner of the board, which is Nihonbashi in Edo. It then goes through in sequence the 53 stations marked by vertical red labels scattered around the map. It finally ends at the destination, Kyoto, at the top right corner of the board, echoing the word "up-capital" in the title. The board is so large that it not only gives beholders an overall view of the vast territory, which is what the word "ichiran" in the title means, but also lets players zoom in the road when moving tokens across the stations, providing a virtual travel experience.[14] The bird's-eye view map of this sugoroku is of course not a faithful geographical map but Hiroshige's imaginative representation of the coastal road modified to fit the board frame (see Figure 7.4). When the path runs across the board on the left and passes by Mount Fuji, it still roughly matches the actual geography. After going around the top left, it meanders downward into the right side of the board where the other side of Mount Fuji is visible from those stations. This cannot be any further from the real-world Tōkaidō. That said, this sugoroku is faithful to the facts in the sense that the boat ride connecting stations 41 and 42 is depicted near the bottom right corner of the board. Another unique feature of this sugoroku is a fork after the 43rd station, Yokkaichi. While the right branch follows the original road to the capital, the left branch is a pilgrimage route leading to Ise Grand Shrine, as informed by the word "pilgrimage" in the title. Players might choose to go to the capital or visit the shrine.

This sugoroku is a more distant variant of the travel narrative of the Hiroshige's ukiyo-e print series. The road scenes presented on the sugoroku board cannot directly correspond to the original ukiyo-e prints, because the former is a continuous bird's-eye view and the latter consists of many varied viewing angles. Yet, the

FIGURE 7.4 A digital archive of *Pilgrimage Up-Capital Dōchū Ichiran Sugoroku*.

Source: National Diet Library's website, https://dl.ndl.go.jp/pid/1310712, accessed 2024–06–26.

changing landscapes of the former do offer some hints. For example, the tenth station, Hakone, on the sugoroku path stands at the top of a hill on the lower left of Mount Fuji on the board. This matches the corresponding print in the series that features lush towering mountains with Mount Fuji in the faraway background. One could imagine taking in a view similar to that in the ukiyo-e print from somewhere in the lower left of the sugoroku board. Other examples include the nineteenth station, Fuchū, which in both the ukiyo-e series and the *Tōkaidō 53 Stages New Board Dōchū Sugoroku* previously mentioned, shows people walking across the river. On the bird's-eye view map of this sugoroku, a river can be found between the nineteenth and twentieth stations. All in all, this sugoroku offers a consistent bird's-eye view, whereas other dōchū sugoroku boards and the original ukiyo-e prints present each station from varied perspectives, but they all show a similar travel narrative on the same terrain.

This bird's-eye view sugoroku can be considered as an analogous story for mapping and blending with the laundry routine too. When folding the clean laundry at home, a person might virtually take a bird's-eye view of the whole Tōkaidō on the sugoroku board and zoom in a station. With the external knowledge frame [accommodation] invoked, and probably another frame [road map] to be brought in, the person can imagine that the sugoroku board shows one's current location on the road and the zoom-in view simulates the walk to the next station. After arriving at the next station, one can imagine staying in another guesthouse and doing the laundry again soon. This sugoroku can be regarded as a road-map variant of the travel narrative on the Tōkaidō.

Famous Places of Edo Ichiran Sugoroku (江戸名所一覧双六)

Pilgrimage Up-Capital Dōchū Ichiran Sugoroku presents a distinctively stylized path embedded in a bird's-eye view map, but it is not the only ichiran sugoroku. *Famous Places of Edo Ichiran Sugoroku* presents a bird's eye view of Edo with vertical red labels marking locations of famous views scattered in the city.[15] This sugoroku board was designed by Utagawa Hiroshige II and published in 1860. The actual size is similar to the other ichiran sugoroku. The starting point is Nihonbashi, at the center of the board, which is also the finishing point.[16] Every vertical red label is numbered, starting from the second station, which is on the immediate left of Nihonbashi. The path then traverses different parts of the map and finally arrives at the starting point from its right. There is no clearly pre-defined path guiding the players from one label to the next, but they can move the tokens through the roads or bridges in the map, which also gives them a sense of a virtual walk in the city. One might also choose to detour rather than walking the shortest path to the next numbered label. Compared with Hiroshige's another famous ukiyo-e print series discussed in Chapter 5, *100 Famous Views of Edo*, this ichiran sugoroku consists of only 55 famous places in Edo, and many places from the two artifacts are different. That said, this ichiran sugoroku can be interpreted as a road-map variant of the ukiyo-e print series.

100 Famous Views of Edo is considered in Chapter 5 as an analogous story for mapping with maintenance routines such as regularly having takeout lunches during workdays. Having a takeout somewhere in a city corresponds to a famous view in the print series. This ichiran sugoroku can instead allow a person performing the routine to check off the list of famous places on the map. When a person buys a takeout meal and walks to a public space nearby to enjoy it, one can invoke the external knowledge frame [public space] and imagine a third story of finding a famous scenic location in Edo. Taking a seat in the public space, one virtually checks off a numbered label, whose position relative to the starting point on the sugoroku board matches the actual location of the public space relative to one's workplace on the real-world map. Frequently having takeout lunches allows one to check off many famous places on the sugoroku board. Since the famous places are evenly distributed throughout Edo city, the person would also be encouraged to explore different public spaces near their workplace in the real world.

New Travel Variants

The above sugoroku boards demonstrate possible variants of the travel narratives originally implicated in the well-known ukiyo-e print series. Each of them presents a similar travel narrative, on the Tōkaidō or in the Edo city, from a varied perspective in a distinct pictorial form, ranging from a photo reel of wide shots on landscapes, roads, or rivers, a long horizontal picture simulating motion parallax, a comic strip that focuses on two particular travelers, and road maps that provide a bird's eye view and allow zoom-in for virtual walks. These varied visual perspectives constitute a toolbox for designing and creating new analogous stories or variants. In case a person doing the laundry is not interested in the roads of the Edo-period Japan, they may consider other roads in the world. This is another strategy drawn from the chain tale variants, namely, looking for real-world references. For instance, Iceland is famous for its natural landscapes including breathtaking waterfalls and everchanging glaciers; most of them can be covered on a two-week road trip on the Ring Road. One can refer to Internet sources and pin all the landmarks along the road, starting from the Harpa concert hall in the capital on the first day, followed by Gullfoss waterfall on the second, Seljalandsfoss waterfall on the third, and so on. With the iconic "postcard" images collected from public domain, one can easily create their very own version of print series with an Icelandic road trip theme. One can also arrange the landmark images on the Ring Road to create an adapted and updated version of sugoroku. Bird's-eye view maps of Iceland also exist, for example, Jean Antoine Posocco's Illustrated Map of Iceland using a cartoon style to show the island with the main road, landscapes, landmarks, villagers, tourists, and animals. A beholder of the map can definitely take a virtual walk along the Ring Road and visit the numbered attractions marked on the map. A person performing a routine may use similar illustrated maps of Iceland for their own version of ichiran sugoroku. After finishing the laundry, they may zoom in the map and virtually walk to the next landmark. The previously invoked knowledge frames [accommodation] and [road map] with the imported relations all apply. The person can imagine a third story of touring by car around Iceland, arriving at a landmark, staying in a nearby guesthouse, doing and packing the laundry, and then heading to the next attraction. In short, the person can plan any road trips in the real world as they fancy and use the toolbox to select a representation style for the blend.

The Gojyusantsugihokei (53 Stages Pedometer) app
(東海道五十三次歩計)

The travel narrative on the Tōkaidō is interestingly found to be an analogous story presented in a mobile app. Gojyusantsugihokei is a Japanese step counter app. Instead of the laundry routine, it maps the travel narrative on the Tōkaidō with actual walking. Typical mobile phones of today are able to track and report users' step counts based on data from their built-in sensors such as accelerometers and gyroscopes.[17] Drawing on reports from the mobile phone, the app not only provides an aggregated summary of step counts and walking time on a daily, weekly, or monthly basis in bar charts but also maps the step counts onto the actual distance from Nihonbashi to Sanjō Ōhashi, the departure and destination of the Tōkaidō. It subtly projects a travel narrative on the road through displaying a geographical map covering a graphical route of the Tōkaidō, with part of the road highlighted in red to mark that part as "done" according to the accumulated step counts (see Figure 7.5). It also shows the next station the user is heading to and displays prominently the percentage of the road completed. The narrative seems to evoke the traditional travel practice in the old days, as the icon for the step counts is a pair of zouri, the traditional Japanese grass straw sandals. The calories burnt are conceptualized in servings of tokoroten, a popular Japanese gelatinous sweet during the Edo period, and are symbolized by a graphic of a traditional tokoroten dispenser. The app attempts to draw a travel narrative on the Tōkaidō in a culturally dated context analogous to the user's daily walking. Compared with the laundry routine, daily walking can be readily mapped and blended with the Tōkaidō travel narrative. The app presents a virtual geographical map of the Tōkaidō to stimulate one's invocation of an external knowledge frame [navigation app] and import of new relations like GUIDE (source, path, goal).

However, one might argue that the technological knowledge frame is not perfectly applicable to the culturally dated travel narrative context that the app attempts to project. Here, the aforementioned sugoroku variants can come into play. For instance, *Tōkaidō Gojūsantsugi Sugoroku* is a stamp-collecting variant of the Tōkaidō travel narrative. When the user's accumulated step counts reach the equivalent distance from the starting point to the first station, the user virtually collects the first station stamp on the sugoroku board in the app. The external knowledge frame [travel stamp] for the blend of this variant is compatible with the cultural context of zouri and tokoroten. Another variant *Tōkaidō 53 Stages New Board Dōchū Sugoroku* simulates high-speed motion parallax transitions from one station to the next, which is less congruent with the culturally dated context. Consider the next variant *Ukiyo Dōchū Hizakurige Comic Sugoroku*. With a certain quantity of steps achieved, the person virtually sees in the app a panel of comic about the two awful protagonists' misadventures at the corresponding station. The blend with this variant requires an imaginary comic artist walking together with the user and documenting the comic scenes, which fit the traditional cultural context. Finally, *Pilgrimage Up-Capital Dōchū Ichiran Sugoroku* can be the most promising match. It is a road-map, or to be specific, bird's-eye view map, variant. When the person accumulates a certain number of steps on a day, the app zooms into the bird's-eye view map and moves slightly on the road according to the distance equivalent to the step count. It simulates the scope of view of a player focusing on the sugoroku board. The person can imagine carrying the game board on a trip and tracing the road with their eyes, after a day's journey. The sugoroku perfectly matches the zouri and tokoroten context.

FIGURE 7.5 A screenshot of the Gojyusantsugihokei app used by the author.

Variants of the Nine-Nines Charts

Travel narratives conveyed by ukiyo-e print series or sugoroku boards, each of which shows a sequence of different scenes resulting from repetitions of a similar walking action, belong to scene-to-scene storytelling. Counting days on a calendar, likewise, produces a sequence of scenes (i.e., records), resulting from repetitions of a similar drawing or writing action. The nine-nines charts, as introduced in Chapter 6, feature distinct cultural narratives of counting the winter days and documenting the bitter weather. The charts have evolved into varied forms since its existence dated back to the Yuan dynasty (ca. 1271–1368) of the Imperial China. The earliest description can be found in a poem and its note by Yang Yunfu of the Yuan dynasty. The form at that time was like a "mixed-media" personal life documentation, which includes a typically picked plum spray pasted on a paper window screen and clusters of red circles drawn on the paper to mimic the flowers. It is believed that individuals of that period used rouge to draw one circle standing for one flower each day. Nine flowers formed a cluster, and nine clusters filled up the spray, indicating the passing of the winter. In the Ming dynasty (ca. 1368–1644) that followed Yuan, the charts continued to evolve, and a physical sample of a varied form has survived today. It includes an ink rubbing kept in National Library of China and the stone in Xi'an of Shaanxi Province.[18] The dimensions of the rubbing are 91 centimeters in length and 41.3 centimeters in width. The rubbing features in its center a spray of plum in a vase bearing nine clusters, each of which comprising nine flowers, for counting the winter days. It seems to be an impression variant of the preceding mixed-media version of the charts. According to its dimensions, the visual representation of the plum spray is about 16 by 13 centimeters in size, which is slightly smaller than the actual size of a spray bearing that number of flowers. The mixed-media version, in comparison, is relatively true to life in scale, because of the physical plum spray. In the Qing dynasty (ca. 1644–1911), the charts transformed from figurative representations to graphical or linguistic representations. The graphical variants typically present a three-by-three matrix with each of the cells containing nine hollow circles for people to fill on each day. Different cells may have the nine circles arranged in different abstract patterns.[19] Different ways of filling a circle indicate the recorded weather of that day. For example, filling the top half for clouds, the bottom for a sunny day, the left half for rain, the right half for wind, and the whole circle for snow.[20] The linguistic variants are discussed at length in Chapter 6. A chart of this kind presents nine characters, each of which is composed of nine strokes, forming a poem. Some poetry charts even have weather descriptions written on individual corresponding strokes. All in all, each variant of the nine-nines charts presents a narrative of someone recording the passing of days. Variants share the same narrative structure, dividing the period into nine groups of nine days, which is the nexus. Meanwhile, the days are presented or represented very differently among variants. There are two levels of variations, respectively, on the two levels of frames mentioned in Chapter 6. On the first level, variations are found among a cluster of flowers, a pattern of hollow circles, or a Chinese character. On the second level, differences are spotted among a spray bearing many flowers, a matrix of circle patterns, or a Chinese poem.

New Daily Logging Variants

Differences in presentations or representational styles, varied from the figurative, the graphically coded, to the linguistic, inform ways of updating, adapting, or personalizing analogous stories through making variants. In Chapter 6, I suggest that the nine-nines poetry charts can be read and used as analogous stories for mapping with maintenance routines like walking the dog every day. If a person is not interested in poetry, they might consider the figurative alternatives. Referencing the nine-nines charts in the Yuan period, they can pick up a fallen spray of plum, stick it on a sheet of paper, and draw a red circle every day. If plum or apricot flowers are not common where the person lives, they might check the local flowers calendar and improvise accordingly. To match the winter period from Dongzhi to March, one local species flowering in winter and another species with similar flower color blossoming in February or March would be the perfect candidates. In Hong Kong, for instance, golden penda is evergreen, and it blossoms from November to February displaying golden yellow flowers, which is followed by yellow pui which flowers from March to April. Both trees are of similar height and their flowers are both yellow.[21] The pairs can substitute the plum and apricot in the original nine-nines charts emerging from northern China during the Yuan dynasty. A Hong Kong resident of today can stick a fallen spray, which can easily be found in the country parks of the city, on a sheet of paper, and draw a yellow circle for a golden penda blossom every day starting from Dongzhi. When spring approaches in March, the drawn yellow circles may represent yellow pui flowers instead. This traditional mixed-media daily practice can be "upgraded" by using digital media and technologies. Consider the person walking the dog in winter. After a walk in December or January, the person comes back home and sees a virtual golden penda tree blossom right before their eyes, probably after a tap on a large-screen display mounted on the wall for true-to-life scale and image quality. With an external knowledge frame [seasonal flowers] invoked, new relations like FALL (flower, road) and PICK UP (flower, road) can be imported to connect the act of walking the dog on the road with the imaginary act of picking up a fallen golden penda on the same road. The person then can blend the real-world walk with the virtual flower on the screen and imagine a third story of walking the dog, bringing a fallen seasonal flower from outside, and sticking on the spray at home. Digital media of today enables the immersive virtual experience through the large-screen display, projection mapping, augmented-reality headsets, or other devices. Information technologies support the real-world connection by pulling up-to-date contextually relevant information like a location-based seasonal flower calendar and dynamically choosing a seasonal flower for the person on each day. Thereby, the analogous story for blending with the routine can continue to unfold not only in different places but also across seasons. For instance, the virtual flowers for someone in Hong Kong can be rhodoleia, with red drooping bell-like flowers in clusters from March to April, followed by scarlet sterculia, with a pale pink star-shaped flower from April to May, and then flame tree with bright scarlet flowers from June to July. The rhodoleia trees are generally shorter (up to 12 meters) than the scarlet sterculia and flame trees (up to 20 meters). The virtual flowers in order from March to July can be put on the virtual spray from lower positions to higher positions. All in all, variants of the nine-nines charts provide templates of different analogous stories to meet different individuals' preferences or needs.

The Wordle Game: A Distant Variant

The nine-nines charts, since the Yuan period, have evolved into varied forms, from the figurative to the linguistic and coded, while the narrative of daily logging remains.

A distant variant of the traditional daily logging narrative can be seen today, which is the web-based word puzzle game Wordle mentioned in Chapter 1. The game once was an online fad in 2021, probably owing to the global lockdown during the COVID-19 pandemic period. Players across the globe attempt to solve a five-letter puzzle on the Wordle web page. Each player has six chances to guess the word, and for each unsuccessful guess, the player receives a color-coded hint wherein a green square means a correct letter in correct spot, a yellow square means a correct letter in wrong spot, and a grey square means an incorrect letter. After the game, the color-coded hints are put together into a pattern of colored squares in five columns and at most six rows. The game looks like another simple word puzzle, but it invites people to play every day with a different word puzzle on a daily basis. It has successfully become a daily practice of many people for a period of time. Players record and even share online their daily results in the form of the color-coded patterns without giving away the solution to the puzzle. A parallel can be drawn between playing Wordle every day and logging the days on the nine-nines charts. First, both the game and the charts require people to perform a simple daily routine, which only takes minutes to finish. Second, both of them show the results of one's routine in some kind of code. Wordle encodes the result of a game in a colored square pattern, while the nine-nines charts in the coded form let users to jot down weather of a day by filling a part or the whole of the corresponding circle. Without referring to the game instruction, or reading the note written next to the chart, one could not decode the result of a game played, or the weather of a previous day. Both playing Wordle and using a nine-nines chart can be interpreted as narratives of daily logging. They can be regarded as distant variants as well, if one sees the nexus element as logging results of daily practice. Nuances between Wordle and the nine-nines charts in the poetry form can be found in the linguistic representation. Wordle provides one five-letter English word every day; a nine-nines poetry chart incrementally shows a Chinese character composed of nine strokes every nine days. The Chinese characters appear in order on the chart and gradually form a poem. The five-letter words released by Wordle daily, on the other hand, do not and even cannot constitute any expressions, whether a poem, a sentence, or a quote, because that may provide unintended hints to the upcoming puzzles, ruining the fun of the game.

Apart from the coded and the linguistic, the nine-nines charts also consist of the figurative form that Wordle has no parallel. The very initial form of the charts began as an actual spray pasted on the paper window with flower-like circles drawn daily, followed by an ink rubbing depicting a spray bearing hollow flowers for one to fill in. The charts are presented figuratively as a plum spray, and every filled flower represents a day. The presentation of Wordle's results is in purely graphical patterns, which are easily shared on social media, partly contributing to the widespread popularity of the game. Meanwhile, some might imagine the colored square patterns being more figurative. Tarmo Annus, a programmer in Estonia, has published a free converter for transforming the colored square patterns into colored houses in Townscaper, a city builder game by Oskar Stålberg.[22] Townscaper allows players to freely build colored, cartoon-like houses one by one in a three-dimensional virtual environment on a distorted grid of sea. One can build a house in a spectrum of rainbow colors including green and yellow. A new floor can go on top of an existing roof or balcony, and lateral extensions are possible on automatically generated stilts. The colored square patterns resulted from Wordle can be mapped, in two dimensions, onto the colored houses in Townscaper, with each Wordle row turned into one level of the building in Townscaper. The grey squares, which indicate incorrect letter in a guess, can be rendered as either gaps or grey blocks in the houses (see Figure 7.6).

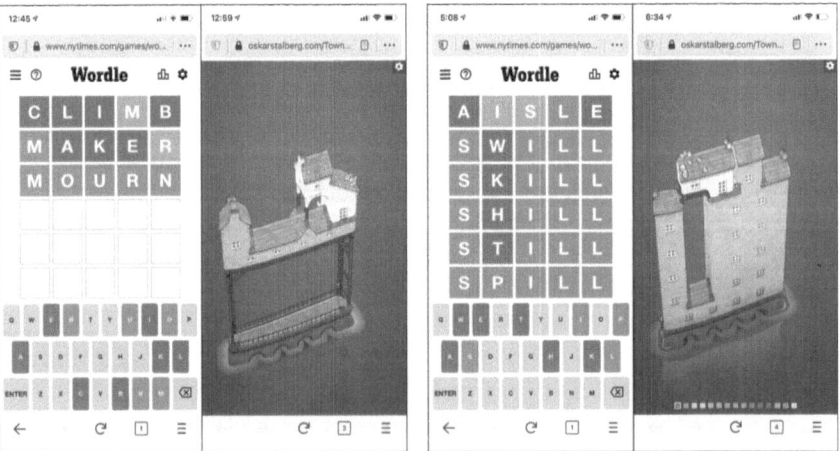

FIGURE 7.6 Screenshots of Wordle and Townscaper results played by the author. Results from playing Wordle can be parsed by a free online tool and converted into a virtual building rendered by Townscaper.

This converter intriguingly enables a new figurative form of the Wordle narrative of daily logging. A player can keep daily records of playing the game not only in the color-coded patterns but also in the form of colored beautiful houses. Every day, one logs a result and builds a new colored house on the three-dimensional virtual sea, gradually forming a beautiful, old-styled town after a period. With ambient animations featured in Townscaper, like flying birds and dynamically changing natural light, one can display the town on a large-screen display and imagine looking through the window for a tranquil view that soothes the nerves. The Wordle's results, in forms of the original color-coded patterns or the transformed colored houses in Townscaper, constitute scene-to-scene narratives of daily logging. They are also distant variants of the traditional nine-nines charts. The Wordle variants can be used as analogous stories for mapping with daily maintenance routines like walking the dog. During the walk, one might take a rest on the road and attempt to solve the word puzzle. After the walk, one can see today's result in the form of a colored house displayed on the screen at home. The knowledge frame [visual journal] originally invoked for the blend of the nine-nines poetry charts still applies here, plus another external knowledge frame [house construction]. One can imagine a third story of virtually seeing a newly built colored house on the road while walking the dog. Looking at the virtual town on the screen after the walk is similar to recapping the colored houses that one has come across on the road over a period of time via a virtual window.

Variants of *Incomplete Open Cubes*

In this chapter, we have looked at variants, which inform us of ways to update, adapt, or personalize analogous stories. Drawing from many variants of folktales, which cover different characters and settings, we gather insights such as real-world references, special social groups, and preferred professions. Variants of travel narratives in the Edo period were spun off in multiple media forms, such as ukiyo-e print series, comical novels, and sugoroku game boards, which can be summarized as different narrative presentations from photo reels, motion parallax transitions, comic strips, and bird's eye views. Variants of the nine-nines charts demonstrate different representations, including the figurative, the coded, and the linguistic. These many variants embody different possibilities of creating variations of a story by changing the content (e.g., characters, settings), the presentation style (e.g., motion parallax, bird's-eye views), or the representation (e.g., figurative, coded). Individuals who practice a routine and find an analogous story uninviting may consider changing any of the above aspects and look for a new variant to one's taste and temperament. Folktales, Edo-period travel narratives, or ancient Chinese nine-nines charts belong to folklore or popular culture, and hence variants are commonplace. In contrast, analogous stories drawn from art, like Sol LeWitt's serial art project as discussed in Chapter 6, are more singular. A work of art is typically an embodiment of an artist's unique thought or sensation. Each work thus should be distinct in attributes like content, form, the motif, or at least the statement. Sometimes, a series of works by the same artist with the same motif are regarded as a series work, such as Monet's *Haystacks*. For Monet, *Haystacks* is by no means the first time he labeled his paintings a "series," but he uniquely identified the series as his attempt to depict "instantaneity" of the "same light diffused over everything."[23] Hence, the artist's statement of *Haystacks* is distinctively mentioned. The narrative of the series can be interpreted as a sequence of depictions of diffused light on a field with haystacks at different instants. If any variants of *Haystacks* exist, while the motif might change, light should be the nexus element common to all, comparable to team building theme among "The Flying Ship" and its variants, the 53 stations on the Tōkaidō among the ukiyo-e print series and sugoroku, or daily logging among the variants of the nine-nines charts. The same logic applies to Sol LeWitt's serial art project, *Incomplete Open Cubes*. As mentioned in Chapter 6, the core of serial art is a systematic iterative process that embodies an artist's premise. The systematic process of *Incomplete Open Cubes* is enumerating all possible cubic structures with one or more edges removed under certain criteria (i.e., rules). The nexus element among variants of the process, if that can be imagined, should be removing edges from a structure. Following the thoughts related to serial art including minimalism, one might imagine variants by changing the structure from cube to other primitive three-dimensional forms like tetrahedron or pyramid, or two-dimensional shapes like a hexagon.

In Chapter 6, I suggest that *Incomplete Open Cubes* can be an analogous story for mapping with routines like dollar cost averaging (DCA), which demands perseverance and long-term engagement on the part of the investor. The minimalist approach of the original project that chooses the wireframe cubes might not be able to engage some people in a long run. A person might prefer structures or patterns that are more contextually or culturally relevant. That can deviate from the original principle of serial art, but the "premise" here is personalizing analogous stories for expressive iteration rather than being faithful to an art movement. That said, the new structure or pattern should still be compatible with the nexus process, which is having edges or segments removed from a basic form. For instance, a regular hexagon (equilateral and equiangular) is a simple two-dimensional shape that can be associated with beehives or some lava stone columns, which are "engineered" by nature. When used as a pattern or an icon, it typically suggests science, technology, engineering, mathematics, or in short, STEM, which seem to be contextually relevant to the DCA routine that is grounded in financial analysis. Meanwhile, a hexagon has only six edges to be removed, which may render too few variations of incomplete hexagons. New variants should also follow the original criteria that the valid pattern should be connected, and two patterns are identical if they can be transformed into each other by rotation. It follows that a hexagon with six diagonals through its center to its six vertices, that is, a dissected hexagon, would provide more variations to be enumerated. Yet, a dissected hexagon, which is composed of six equilateral triangles, loses its visual resemblance to beehives or lava stone

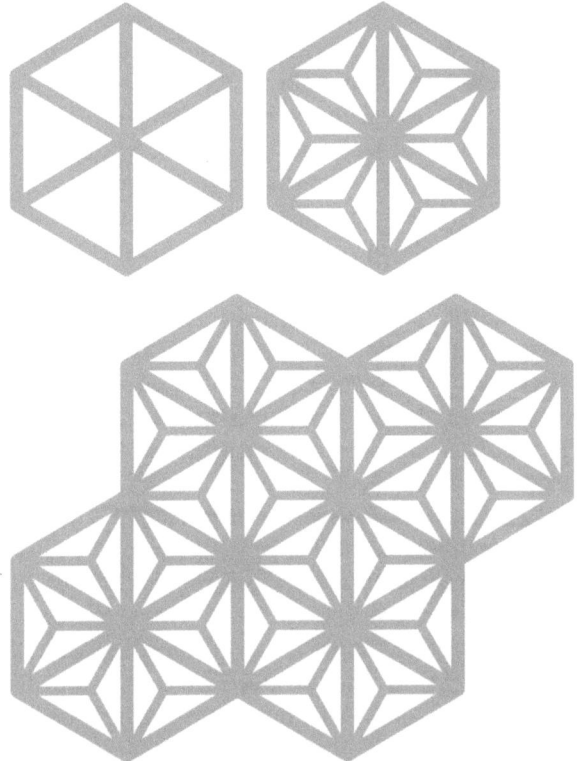

FIGURE 7.7   Illustrations elaborating from a regular, dissected hexagon (top left), to an asanoha pattern (top right) after segments added, to a patch of asanoha patterns (bottom) resembling a hemp field.

columns and reverts to an abstract shape (see Figure 7.7, which coincidentally is the same shape as the isometric projection of a cube presented in Chapter 6). To further explore the representational potential of the shape, one might add extra segments inside the equilateral triangles, which gradually constitutes a culturally resonating pattern, called "asanoha," a traditional Japanese pattern symbolizing hemp leaves (Figure 7.7). Hemp is a fast-growing plant with strong fibers that can be used for producing textiles, paper, rope, and clothing, among others. Elaborated on dissected hexagons, asanoha resembles hemp leaves, and repetition of the pattern looks like a field full of hemp. In Japan, the pattern symbolizes "auspiciousness." It is traditionally used in babies' and children's clothing with the hope that they could grow fast and strong.[24] This cultural meaning of asanoha renders it a good match with the DCA routine in terms of growth.

Asanoha is also one of the common patterns used in kumiko, Japanese woodworking that assembles thin wood bars of different colors into intricate patterns with precise angles, tongues, and grooves. The pieces hold together and remain in place just by mutual pressure without the use of any nails. The sophisticated art and craft are used to make human-sized panels in interior design or in smaller scale for decorating household furniture like closets.[25] I once received a kumiko assembling kit of asanoha hexagon as a souvenir.[26] It provides pre-cut, ready-to-assemble wooden pieces and clear step-by-step captioned graphic instructions for one to follow. The intermediate outcome is a dissected hexagon frame. One then puts three short pieces inside each equilateral triangle, with a total of 18 pieces, resulting in a hemp leaf pattern. The assembling of asanoha kumiko seems to be a traditional art counterpart of the iterative process of constructing incomplete open cubes. The former inspires a new variant, Incomplete Asanoha, for mapping with the DCA routine. The variant starts with a regular, dissected hexagon. When the person makes the first investment, they receive one virtual rod to be put inside an equilateral triangle of the hexagon, resulting in the first incomplete asanoha, with 17 segments removed (or only one segment added). This serves as proof-of-work for validating the purchase the person has just made. The external knowledge frame [blockchain technology] originally invoked still work for the blend of this new variant with the routine. After the second investment, the person receives the dissected hexagon (without any segments) again and another virtual rod. According to the rules of the serial art project, two patterns are identical if they can be rotated into each other, but not so by mirror reflection. There should be three variations of incomplete open asanoha with 17 segments removed (see Figure 7.8). The second and third variations should be easy to find. From the fourth purchase onward, the person receives two virtual rods to be put inside the dissected hexagon to come up with an incomplete asanoha with 16 segments removed. This degree of incompleteness (16 segments removed) requires more cognitive efforts to enumerate all the variations. Anyhow, every time a new variation of a certain degree of incompleteness is cataloged, it marks a regular purchase. The varied incomplete hemp leaves found are put together one by one, forming a large virtual kumiko panel resembling a hemp field. This virtual kumiko panel can be projected on the wall or shown on a large digital display, like its physical counterpart used as a motif on traditional sliding doors, feature walls, or ceilings, that surrounds an interior space with hopes of growth.

FIGURE 7.8   Variations of incomplete asanoha with only one segment (i.e., 17 segments removed)

Summing Up

Different routines require different analogous stories to match. Since routines involve repeating actions, analogous stories should be composed of recurring events. In goal-directed routines, people are concerned with progress. Chain tales feature simple chains of recurring events that grow in size or pile up items, which can be mapped with routine acts to indicate the progress made. Different categories of chain tales match routines in pursuit of different goals. Chain tales of plot types "the quest" and "overcoming the monster" work for definite goals, those of plot type "voyage and return" are perfect for checklist goals, and lastly those of plot type "rags to riches" fit long-term goals. In maintenance routines, variety is preferred to progress. Theme-based stories like ukiyo-e travel narratives present series of varied outcomes instead of repeating actions, which can add color to the mapped routine acts. Meanwhile, a routine can last longer than the iteration of events in a chain tale or a theme-based narrative. Hence, iteration of an analogous story into sequels and then an episodic or anthology series spun off from the original story can better match a long-running routine in real life. Between analogous stories in a series, people sometimes take a break and stop the routine. To enhance continuity across the series and to encourage people to resume the routine, adding a frame story to envelop all theme-based narratives or weaving an overarching storyline through all chain tales are two approaches for consideration.

Yet, people might suspend or quit a routine due to many reasons. In case a person wants to restart the routine after a hiatus, one might not remember the previous analogous story. Sometimes, one just gets bored of the repeating theme or characters of the story. It is time to "reboot" the story too. Here, variants come into iteration. First, chain tales for different goal-directed routines commonly exist with their variants, which inform approaches to updating or personalizing an analogous story, namely, looking for real-world references, considering different social groups, and identifying preferred professions. Second, variants of ukiyo-e travel narratives suggest different presentation styles, such as a photo reel, a horizontal picture with motion parallax transitions, a comic strip, and a bird's-eye view map. These varied presentations offer templates for adapting an analogous story in a changing context. The nine-nines charts that feature daily logging narratives, which are theme-based stories in multiple frames, also see many variants. They varied from the figurative, the linguistic, and the graphically coded forms, which suggest possible representations that can transform an analogous story. Lastly, Sol LeWitt's serial art project is taken to generate a new variant for the long-term investing routine, with reference to the aforementioned guidelines including real-world reference and figurative representation. In this chapter, not only do I describe the existing variants of analogous stories but also demonstrate the creation of new possible variants to one's specific needs, tastes, or situations. Figure 7.9 summarizes the insights from existing variants and distributes the new variants along the continuum between variety-focused and progress-focused routines.

variety-focused process	maintenance routines	goal-directed routines			progress-focused process
		long-term goals	checklist goals	definite goals	
theme-based variants					chain tales variants
↓	The "redesigned" Gojyusantsugihokei app				↓
photo reel					social groups
comic strip				The Fallen Tree	
motion parallax	Iceland Ichiran Sugoroku				
bird's-eye view map					
linguistic	Nine-nines poetry charts		How Eleven Made Their Way in the League		preferred professions
graphically coded	Wordle	The Chef			
figurative	Nine-nines seasonal flowers charts	Incomplete Asanoha			real-world references
	Wordle-to-Townscaper				

FIGURE 7.9   Insights from existing variants (at both ends) and distribution of new variants along the continuum between variety-focused and progress-focused routines.

PART III

Design

The three chapters of Part I, drawing on the concept of rhetoric, the psychology of behavior, imagination, and fiction, as well as the cognitive science of analogy and blending, characterize expression iteration. In Part II, three levels of expressive iteration are examined and discussed. Iteration of causally linked or thematically related events results in two kinds of stories, namely, chain tales or theme-based narratives, analogous to goal-directed or maintenance routines, respectively. The former focuses more on progress, while one prefers variety in the latter. Aligning and blending an analogous story with a matching routine

TABLE III.1 A summary of the exemplary cases of analogous stories demonstrating the first level of expressive iteration, that is, iteration of events, with the matching routines and invoked knowledge frames

Chain tales	Goal-directed routines	Knowledge frames
The Turnip	Regular workouts	Urban farming
The Three Little Pigs	Taking medications	Neighborhood
The Flying Ship	Attending regular classes	Team project
Tsarevitch Ivan, the Firebird and the Gray Wolf	Finishing piecework	Mission titles, named trains
The Stonecutter	Regularly trading used items	Chain reaction

Theme-based print series	Maintenance routines	Knowledge frames
36 Views of Mount Fuji	Regularly taking stairs	Observation tower
100 Famous Views of Edo	Having a takeout lunch	Public space
53 Stages of the Tōkaidō	Doing laundry	Accommodation

DOI: 10.4324/9781003391449-10

yield a third story where an invoked knowledge frame binds together both fictional and real-world elements that are most relevant to a person. Table III.1 summarizes the cases of analogous stories with their matching routines and invoked knowledge frames.

On the second level, an analogous story is iterated into sequels of a similar plot, forming a series. An episodic series of chain tales or an anthology series of theme-based narratives is able to include more cycles that can be mapped with typically long-running routines. I demonstrate the iteration of above cases into different types of series accordingly. Continuity across each episodic series is enhanced by weaving an overarching storyline, while that across each anthology series is strengthened by adding a frame story. The corresponding routines and invoked knowledge frames are summarized in Table III.2. Most sequels share the same knowledge frames with the original stories for blending. Meanwhile, the overarching storylines or frame stories require additional knowledge frames.

Iteration of variants is required in case a person wants a reboot, because of either a lack of interest in the analogous story or difficulty in recalling the original story after a hiatus. Variants can be found in quite many folktales. Sugoroku game boards, a form of Japanese woodblock prints different from ukiyo-e print series, but showing similar travel narratives, are one of such variants. Existing variants inform users of approaches to generating ideas of new variants, which can be mapped with varied routines as far as appropriate knowledge frames are invoked. Table III.3 summarizes the cases.

Through a number of case studies in Part II, I successfully demonstrate the mapping and blending with a few common goal-directed routines like regular exercises, attending classes, and regular trading or investments, as well as maintenance routines like doing the laundry and walking the dog. Yet, these are just part of a long list of what we might regularly practice. In Part III, I attempt to enumerate an extensive list of routines and provide comparisons of different possible analogous stories. In the final chapter, I will discuss how the blends between analogous

TABLE III.2 A summary of episodic series and anthology series developed from the exemplary cases of chain tales and theme-based pictorial narratives, respectively, as well as the overarching storylines and frame stories applied to enhance continuity between stories

New episodic series	*Goal-directed routines*	*Knowledge frames*
The Flying Ship and other Companions	Attending regular classes	Team project
The Stonecutter and more Wishes	Regularly trading used items	Chain reaction

(Continued)

TABLE III.2 (Continued)

New episodic series plus overarching storylines	Goal-directed routines	Knowledge frames
The Turnip and other Root Vegetables Overarching: a continuing quest of the grandfather	Regular workouts	Urban farming Self-determined goal
The Three Little Pigs and other Threats Overarching: an enduring monster facing the pigs	Taking medications	Neighborhood Defense system
Finding variations of: - all three-edge open cubes - all four-edge open cubes ... - all 11-edge open cubes Overarching: cataloging variations of *Incomplete Open Cubes* as proof-of-work for successive transactions	Dollar cost averaging	Blockchain technology

New anthology series within frame stories	Maintenance routines	Knowledge frames
Famous Views of the 60-odd Provinces *100 Famous Views of Edo* Frame: hosting a competition of visual travelogues and introducing each entry	Having a takeout lunch	Public space Public space Contest
69 Stages of the Kisokaidō *53 Stages of the Tōkaidō* Frame: self-entertaining and researching by reading travel stories while waiting for actual travel	Doing laundry	Accommodation Accommodation Reading
Daily logging by writing Chinese characters to sketch out things seen on the path - 春 ... - 室 1st level frame: completing a poem as caption for the sketches 2nd level frame: counting days using the chart	Walking the dog	Visual journal Calendar

stories and target routines can be simulated by emergent and accessible technologies like virtual reality (VR), augmented reality (AR), or mixed reality (MR), and generative artificial intelligence (AI). Lastly, I shall conclude by suggesting iterative expression as a promising way to embody the identity of humanity in response to the rise of AI.

TABLE III.3 A summary of existing and new variants of different analogous stories together with their target routines and suggested knowledge frames

Variants of chain tales	Goal-directed routines	Knowledge frames
The Turnip	Regular workouts	Urban farming
The Fallen Tree (new)	Regular workouts	City trails
The Three Little Pigs	Taking medications	Neighborhood
The Three Little Pigs: An Architectural Tale	Taking medications	Neighborhood
The Flying Ship	Attending regular classes	Team project
The Six Servants	Attending regular classes	Team project
How Six Made Their Way in the World	Attending regular classes	Team project
How Eleven Made Their Way in the League (new)	Attending regular classes	Team project
The Stonecutter	Regularly trading used items	Chain reaction
The Fisherman and His Wife	Regularly trading used items	Chain reaction
The Chef (new)	Regularly trading used items	Chain reaction

Variants of travel narratives	Maintenance routines	Knowledge frames
100 Famous Views of Edo	Having a takeout lunch	Public space
Famous Places of Edo Ichiran Sugoroku	Having a takeout lunch	Public space Road map
53 Stages of the Tōkaidō	Doing laundry	Accommodation
Tōkaidō 53 Stages Sugoroku	Doing laundry	Accommodation Travel stamp
Tōkaidō 53 Stages New Board Dōchū Sugoroku	Doing laundry	Accommodation Motion parallax
Ukiyo Dōchū Hizakurige Comic Sugoroku	Doing laundry	Accommodation Comic artist
Pilgrimage Up-Capital Dōchū Ichiran Sugoroku	Doing laundry	Accommodation Road map
Iceland Ichiran Sugoroku (new)	Doing laundry	Accommodation Road map
The 53 Stages Pedometer app	Daily walking	Navigation app (original) Travel stamp (suggested) Road map (suggested)

Variants of daily logging	Maintenance routines	Knowledge frames
Nine-nines poetry charts	Walking the dog	Visual journal Calendar
Nine-nines mixed-media charts	Walking the dog	Plum and apricot Calendar
Nine-nines digital-media charts (new)	Walking the dog	Seasonal flowers Calendar

(*Continued*)

TABLE III.3 (Continued)

Appropriated Wordle-to-Townscaper (new)	Walking the dog	Visual journal (suggested) House construction (suggested)
Variants of Incomplete Open Cubes	*Goal-directed routines*	*Knowledge frames*
Incomplete Asanoha (new)	Dollar cost averaging	Blockchain technology Asanoha (hemp) pattern Kumiko panel

8

CRAFTING EXPRESSIVE ROUTINES

People in different situations perform different routines for different reasons. A person may practice a running routine to prepare for an upcoming marathon or work out regularly to build muscles. A parent may ask a kid every day and night to take medications or supplements to boost their immunity. A student intends to attend all classes of a language course to prepare for overseas study. A worker commutes every day to finish piecemeal jobs. A person starts a regular fixed-amount investment plan for long-term growth. The reasons behind, like physical fitness, personal or professional development, or growth in wealth, look slightly abstract. Some of them can be specified in terms of definite, measurable goals, such as finishing a marathon within three hours and a half (what is called "sub 3:30"), achieving an optimal body weight, or recovering from an influenza. Some reasons can be transformed into checklists, like attending all the classes or finishing all the delivery orders. Other reasons, like the growth of wealth, are less about reaching a definite goal or completing a checklist, but rather making incremental developments in the long run, which can be regarded as a long-term goal. These routines are goal-directed, as the reasons behind them are instantiated by the person as definite goals, checklist goals, or long-term goals. In performing a goal-directed routine, one is concerned with making progress toward the goal. On the other hand, some routines are performed for reasons that are less related to goals, such as taking a shower every day for personal hygiene and doing the laundry or other chores regularly for household management. A parent may ask a kid to tidy up their schoolbag the night before a school day to instill self-management concepts in the kid. A person of the working class may consume takeout or packed lunches on most working days for convenience. Focusing on desk work entails prolonged sitting and thus needs regular breaks to avoid a sedentary lifestyle. Someone might prefer taking the stairs to using an elevator whenever possible for physical health or

DOI: 10.4324/9781003391449-11

environmental sustainability. One might walk the dog every evening for the health of both the owner and the pet as well as to build pet–owner relationship, set a daily reading session for mental stimulation, and do daily meditation for mindfulness. Some people might need to self-monitor vital signs like peak expiratory flow or blood glucose level due to a medical condition. These are what I call maintenance routines, for people practice them in order to maintain their daily life rather than achieving any goals. One might argue that the pursuit and maintenance of a lifestyle is also a long-term goal. In fact, it can be hard to draw a clear line between maintenance routines and routines directed toward long-term goals like trading used items for sustainability. Some maintenance routines and goal-directed routines are motivated by similar reasons too, say physical health. The nuance here lies between variety and progression. If the person practicing a routine is less concerned with making progress than seeing variations, it is a maintenance routine. Conversely, if one prefers seeing progress more than variations, the routine is a goal-directed one. A categorization of routines can thus span a continuum between the variety-focused and the progress-focused. At the variety-focused end, lie maintenance routines. Moving toward the other side sees routines in pursuit of long-term goals, followed by routines performed for checklist goals. Routines driven by definite goals sit at the progress-focused end. Figure 8.1 illustrates the distribution of varied reasons for routines over this continuum.

Primary reasons behind routines are typically for certain purposes or benefits, like being physically fit for a marathon, getting used to communicating in a foreign language for living abroad, wealth growth, environmental sustainability, or mental wellness. Yet, these objectives or benefits often need quite some time to materialize visibly. As many cognitive scientists or writers have pointed out, people sometimes need "secondary reasons" to continue with a routine, including immediate and intrinsic rewards like natural consequences of the routine acts, novel or unexpected outcomes for the dopamine, as well as satisfying successive outcomes that help one get closer to their vision, that is, the primary reasons (see Chapter 1).

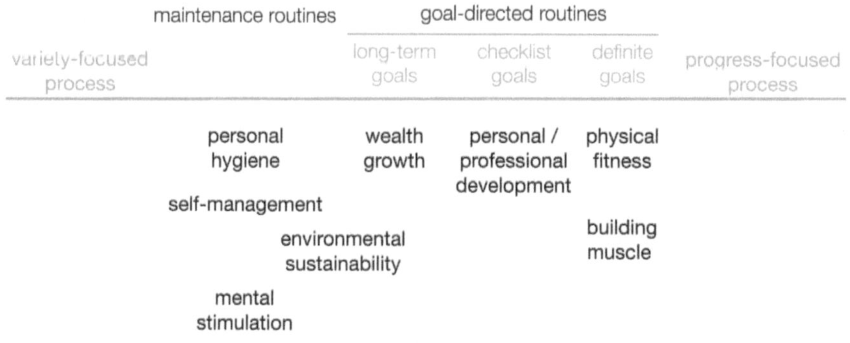

	maintenance routines	goal-directed routines			
variety-focused process		long-term goals	checklist goals	definite goals	progress-focused process
	personal hygiene	wealth growth	personal / professional development	physical fitness	
	self-management				
		environmental sustainability		building muscle	
	mental stimulation				

FIGURE 8.1   Distribution of varied reasons for routines across the continuum between variety-focused and progress-focused processes of iteration.

Applying expressive iteration to routines helps create and suggest new secondary reasons, by means of aligning and blending routines with analogous stories. Different routines motivated by different reasons in different situations require different analogous stories for the blends. Discovering and identifying an appropriate analogous story for a routine follows the well-recognized divergence-convergence design process.[1] First, we look for potentially matching stories whose recurring events are comparable to the repetitive routine acts. This step includes structural alignment between each fictional event and each routine cycle in terms of relations (usually actions) and arguments (usually agents or objects). At this point, a list of candidates would be generated and thus we diverge. Second, we perform blending between each potential analogous story and the routine into a third story. This step involves invocation of a knowledge frame usually outside of both the story and the routine, together with the import of new relations from the frame that can reorganize the components from each side and bind them together in a possible scenario. With the invoked knowledge frame, this possible scenario can be elaborated into succeeding blended events, leading to an ending of the third story. An appropriate ending should be in line with the person's vision, for instance, the original primary reasons behind the routine. Among the candidates, only those with a resonant ending remain. This is a convergence process. Figure 8.2 illustrates the divergence and convergence in the process.

From 2017 to 2022, I conducted a series of design workshops on topics related to story-based motivation for lifestyle behaviors. I provided in the workshops initial design guidelines facilitating the divergence process. The workshops resulted in a collection of story ideas. From 2022 to 2024, I invited expert participants, including educators and practitioners from the creative industries, to independently assess each story idea against others in the collection on varied aspects including perceived causality between routine acts and fictional outcomes, as well as

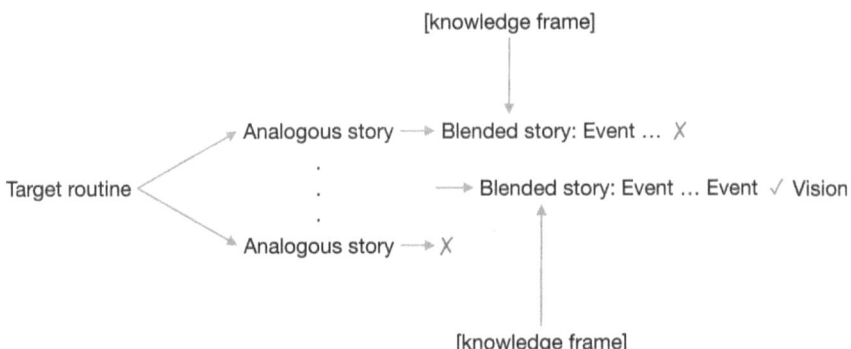

FIGURE 8.2 The divergence process from a target routine, through structural alignments, to yield a list of analogous stories, followed by the convergence process of accessing or learning knowledge frames and elaborating blends toward a story ending resonant with one's vision.

perceived usefulness of the proposed mixed-reality application for the target group. Initial findings suggest that perceived causality and usefulness correlate. Perceived causality can be explained by general accessibility of external knowledge frames for the blends, while usefulness can be associated with alignment of the blended third story ending and the target user's vision. The expert evaluation results seem to support the intended convergence process.

In this chapter, I select representative story ideas from the highly rated, moderately rated, and lowly rated groups in the expert evaluation, and juxtapose them with the cases discussed in Parts I and II. They form a fairly extensive list of typical routines that span the above continuum (Figure 8.1). Each routine can be marked with the primary reasons behind it, which informs the person's vision. And corresponding analogous and blended stories are labeled with suggested knowledge frames and possible endings. The comparison of the vision with the endings demonstrates the convergence process.

The Divergence Process

Technologies have been used in tracking routines. Data are relentlessly generated, extracted, and aggregated around our behaviors via various sources. In a running, workout, or stair-climbing routine, a wearable device like a smartwatch or a smart bracelet can continuingly record a person's vital signs and performance in terms of heart rate, step count, distance covered, flights of stairs climbed, and calories burnt. Location-sensing technologies including the Global Positioning System (GPS), mobile networks, Bluetooth Low Energy (BLE)-enabled iBeacon, or near-field communication (NFC) enable a person to check-in at a place when enjoying a takeout meal or walking the dog in a park, attending a class in the community center, or conducting piecework on-site. Smart IoT (Internet of Things) products like a smart water bottle, a smart pill dispenser,[2] or a smart ashtray[3] can track a user's regular intake of water or medications, or consumption of tobacco. Electronic payment systems like debit cards, digital wallet apps, or online banking can record and report a person's commute as they pay their fare for a train or bus ride, trading used items, or regularly purchasing stocks. Some of these technologies present aggregated information back to individual users in typical data visualizations, like bar charts, line graphs, or colored heat maps. Yet, people do not always act like rational data analysts. In daily routines, general users prefer a more intuitive representation of the information, and even better, an image with emotional attachments and personal relevance.[4]

In a paper of mine,[5] I argue that such a representation or image should be built on a story comparable to the behavior. That means they should share experientially informed similarities in the actions taken, the happenings perceived, the objects involved, as well as spatial or causal relations among them. When the behavior is repeated as a routine, so are the comparable story events, which are virtually presented to the user. With the experiential similarities, the repeated co-occurrence of the behavioral acts and the story events gradually strengthen their connection in the mind-brain, as informed by the embodied cognition thesis.[6] In other words, each virtually presented story event seems to be an intrinsic part of the routine act. The paper proposes a framework for crafting what I call "animated parables" or "mixed-reality parables," a kind of data-driven interactive narrative blended in daily life including personal routines and environments. The framework includes initial design guidelines for crafting a parable of this kind, which consist of several steps. First, look for a simple story composed of cause-and-effect scenarios from a different domain. The causal relations in the scenarios should be direct. Second, the causal actions should be comparable to the behavior in terms of sensorimotor and spatiotemporal experiences, including the visual, auditory, tactile, kinesthetic, motion path, physical contact, proximity, container, and repetition, which are drawn from the basic "image schemas" of embodied experiences, meaning the recurrent structures or patterns in our bodily experiences.[7] Similar sensory images and common spatial or temporal relations between the two sides are noted. Third, the behavior and the comparable story are blended into an interactive narrative wherein different real-life actions lead to different story outcomes. The story outcomes are blended into real-world environments that people encounter in daily routines. Fourth, physical mock-ups, interface wireframes, demo videos, or digital prototypes can be built to visualize the blended mixed-reality scenarios for evaluation. The initial guidelines emphasize exploration in different domains to look for comparable stories, similarities in embodied experiences behind actions, happenings, and objects, structural alignment of similar aspects, as well as selective projection of behavioral actions and story outcomes to be used in the blends. The outlined steps correspond to the divergent part of the design process, which aims to identify more candidates potentially matching a target routine. Hence, invocation of external knowledge frames for pattern completion in blending processes has not been covered yet. This will be addressed in the convergent part.

Since 2017, I have been introducing the initial guidelines and conducting a series of design workshops in my courses on topics related to story-based motivation for lifestyle behaviors. Every two or three students formed a group. In the first week, each group proposed a target behavior and then conducted preliminary research to understand the contexts, including the environments or objects for virtual entities from the story to blend in and possible behavioral data to be tracked by technology. Students then followed the guidelines, generated initial ideas, and presented the mapping and blending. Until 2022, six similar workshops were completed, resulting in a collection of 104 story ideas for a wide range of different lifestyle behaviors. Among the collection, the large clusters include ideas related to personal hygiene (e.g., brushing teeth, showering), working in a sitting position, eating habits (e.g., healthy diets, mindful eating), drinking habits (e.g., drinking adequate water, drinking too much alcohol), and consumption (e.g., using paper, purchase of goods). Other clusters are about exercise (e.g., stretching, breathing), doing chores (e.g., laundry, meal preparation), commuting (e.g., by train, by bike), organizing (e.g., the fridge, the schoolbag), and self-monitoring of bodily conditions (e.g., peak expiratory flow, blood glucose level). While some behaviors are bad habits rather than routines, the broad categories provide references of possible routines set among different people.

Meanwhile, the collection of 104 stories also spans a variety of genres and different themes. They include the so-called "slice of life," that is, common scenes in everyday life, such as playing music instrument or traveling on a train, "animal tales," including a hen hatching eggs and a butterfly in metamorphosis, "fantasy," "adventure," and "science fiction" like feeding the monster pet, saving animals trapped in snow, and urbanizing a planet. The distribution of ideas over different genres and themes shows that participants' imagination is not limited to associations with their own real-life experiences but influenced and stimulated by other fact-based or fictional narratives, probably from novels, movies, or other media (e.g., scouting for truffles, stepping on cracked ice).

Over the period of collecting the story ideas, I repeatedly conducted analyses on the ideas regarding their similarities with the target behaviors in terms of sensory images and image schemas. Sensory images cover typical modalities of perception. For instance, tactile images come from the sense of touch, like warmth or pain; kinesthetic images refer to the internal sense of posture or bodily movement, such as holding something in hand, or blowing air through one's own mouth. Regarding image schemas, I examined the spatial structures or temporal patterns observed in the behavioral acts in the collection. Spatial image schemas include source-path-goal (e.g., climbing "from" the ground "to" the upper floor), above/on (e.g., sitting "on" a chair; rice "on" the dish), container (e.g., putting clothes "into" the washing machine; cigarette ash "in" the ashtray), contact (e.g., direct contact between a needle and the body), and proximity (e.g., smoke "near" the window). Temporal image schemas include repetitive patterns (e.g., watering the plant every other day; stopping at a train station every hour). The examination was first performed idea by idea, followed by focusing on aspect by aspect for consistent interpretations among ideas. The above processes were iterated. The ideas were ranked according to the number of commonalities across all aspects. Those ranked higher have more similarities in physical experiences between the target behavior and the analogous story.

The Convergence Process

From 2017 to 2022, the design workshops resulted in a collection of 104 story ideas for different lifestyle behaviors. The ideas were ranked according to the number of observed similarities. Among those having more similarities between the target behavior and the analogous story, I selected 35 ideas for the next step: expert evaluation. In the previous process, I was the main researcher to conduct the assessments and interpret the observations. In the following step, other relevant domain experts were involved to evaluate the more promising ideas in the collection and separately rate each idea against others with regard to several criteria, including perceived causality, novelty, and usefulness.

Aligning and blending a routine with an analogous story generate and suggest new reasons for one to practice the routine. According to the science of habit as well as positive psychology mentioned in Chapter 1, analogous story outcomes should seem like "natural consequences," "intrinsic rewards," or even an integral part of the routine act. Hence, perceived causality between the routine act and the virtually presented fictional events is crucial. Assessing perceived causality of each idea is necessary. In addition, the story outcomes are suggested to be unexpected or novel. Rating the novelty of the blend between the routine and the story is also recommended. Meanwhile, the story should unfold along a direction that is in line with one's vision, or the primary reasons for performing the routine. But it all comes down to whether the person sees the benefits of using the story. Therefore, usefulness should also be evaluated.

Novelty and usefulness are regarded as two major criteria of creativity. According to the sociocultural definition of creativity, a creative product is one that is "novel" on one hand, and "appropriate, useful, and valuable" on the other. The assessment typically relies on the "suitably knowledgeable social group."[8] The most widely used evaluation method is the Consensual Assessment Technique (CAT) wherein each product in a specific set is rated against the whole set by two or more domain experts independently, and the average rating is a measure of the product's creativity.[9] Studies have shown that the ratings of experts highly correlate in general.[10] In the convergence process of identifying appropriate analogous stories for different routines, I employed CAT and invited a group of relevant experts to rate each story idea's novelty and usefulness, in addition to perceived causality.

Margaret Boden (2009) differentiates two types of novelty. A psychological novelty is new to the person who generated it, while a historical novelty is new in history. The validation of a historical novelty requires a very comprehensive collection of all relevant cases that have existed in the world, followed by thorough comparison against each of them. In contrast, psychological novelty can be evaluated individually. While psychological novelty might include many mundane cases, it is important to note that historical novelty often starts with a talented person's psychological novelty. In the case of evaluating the collection of analogous stories for blending with routines, the relevant domain experts should be knowledgeable enough to assess each idea. If an idea is new to a domain expert, it is psychologically novel and probably also historically novel. If the idea is new to a whole group of experts, it is likely to be historically novel.

Usefulness means not only that a system effectively meets the user's wants or needs, but also that it is easy or convenient to use. Usefulness can be assessed via user evaluation on minimum viable prototypes. In the early stages of the design processes, while prototypes are still rudimentary, perceived usefulness can be assessed. Fred Davis (1989) defines "perceived usefulness" as the "degree to which a person believes that using a particular system would enhance their job performance." To relevant domain experts like designers or design educators, who are trained to empathizing with target users,[11] perceived usefulness can be regarded as how far one perceives a proposed system or product to be useful for what the target users want or need to do. If a group of experts independently evaluate the collection of analogous stories regarding usefulness and reach consensus, the results may cast light on some implicit, unspoken criteria required in the field.

From 2022 to 2024, I successively invited 16 domain experts from two East Asian cities, respectively. They were asked to self-declare their professions and domain-relevant work experiences. The recruited participants were definitely qualified as domain experts in creative practices. As the rater participants were creative practitioners or art and design educators, each idea was professionally visualized on a web-based system in three superimposed panels of scroll-based interactive animations. When selecting the left panel and scrolling down the page, the animation (in a three-dimensional computer graphics style to mimic the real-life experience) synchronously played out the target behavioral act. Selecting the right panel and scrolling down, the rater would see a synchronous animation (in a two-dimensional cartoon style to indicate the virtual experience) of the analogous story. The middle panel corresponded to the blend of the left and the right, that is, the third story. Figure 8.3 shows an example of the three panels. To rate the novelty, one responded on a scale of 1–5 ranging from "not novel at all" (1 point) to "extremely novel" (5 points). For usefulness, the options varied from "not useful at all" (1 point) to "extremely useful" (5 points). For perceived causality, the rater was asked to indicate the level of agreement, from "not at all" (1 point) to "absolutely" (5 points), to

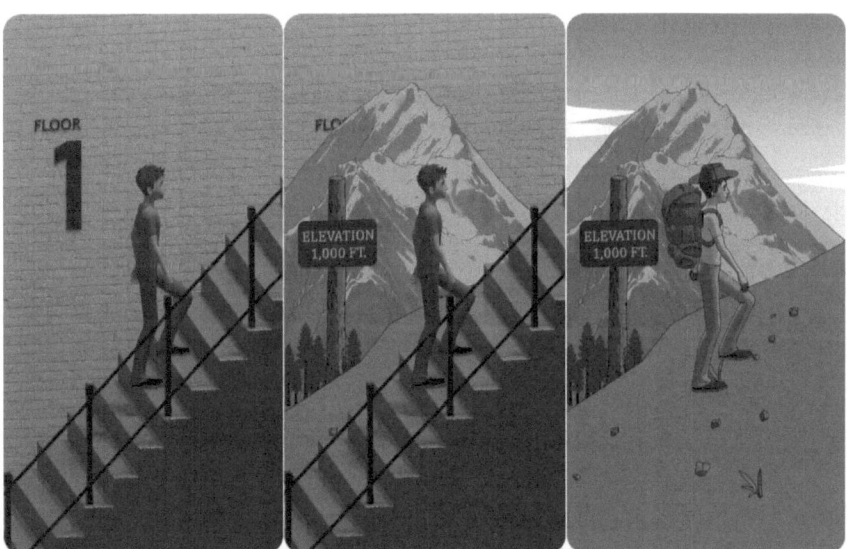

FIGURE 8.3 Screenshots of the three panels showing the routine (left), the analogous story (right), and the blended third story (middle).

a statement like "I feel that" the behavior (i.e., the left panel) "causes," "yields," or "leads to" the outcomes seen in the story (i.e., the right panel). In short, an idea that scored higher points was rated more novel, more useful, or giving a stronger sense of causality between the behavior and the fictional outcomes.

Scores of each idea rated by different experts were collected and analyzed. When employing the CAT, one major criterion is that ratings of the group of experts should highly correlate, that is, reaching certain level of consensus in their assessments. This can be measured by the intraclass correlation coefficient (ICC), which describes how strongly units in the same group resemble each other.[12] We used SPSS to compute the ICCs of the ratings on the 35 ideas regarding perceived causality, usefulness, and novelty, respectively. Among the 16 experts, the ICC of their ratings on perceived causality reached 0.772 (two-way mixed, absolute agreement), which indicated high agreement. The ICC of usefulness reached 0.717 (two-way mixed, absolute agreement), which showed moderate and approaching high agreement. The ICC of novelty unfortunately remained very low. This informed that the experts failed to reach consensus on what constituted novelty.

With the high and approaching high agreements among the experts, the average scores regarding perceived causality and usefulness of each idea are statistically significant and informative. It is found that ideas consensually rated high in perceived causality significantly overlap with those consensually rated high in usefulness. The eight ideas scoring highest on average in usefulness (≥ 3.5) also scored very high on average in perceived causality (≥ 3.44). Meanwhile, ideas rated low in perceived causality (≤ 2.5) and those rated low in usefulness (≤ 2.5) also coincide. This suggests that perceived causality is necessary for an idea to be perceived as useful by the experts. Tables 8.1 and 8.2 list the eight ideas rated high in both usefulness and perceived causality and the five ideas rated low in both aspects, with their working titles, the target behaviors, and the analogous stories.

The expert evaluation identifies eight analogous stories that are rated high in both usefulness and perceived causality. Perceived causality hinges on whether one

TABLE 8.1 Ideas rated high by domain experts in both usefulness and perceived causality

Working title	Target behavior	Analogous story
Pond and Pipes	Turning off the running water tap	Closing the valve to stop draining
Feed Your Bag	Organizing the schoolbag	Feeding the monster pet appropriately
Water Plant	Drinking adequate water	Watering and growing the plant
Mood Bubbles	Practicing breathing exercise	Blowing beautiful bubbles
Sitravel	Focusing on desk work with breaks	Traveling on a train to visit attractions
Peak Flow Flute	Self-monitoring peak flow	Practicing the flute
Staircase Hiking	Taking stairs	Hiking to enjoy spectacular views
Stretching Wings	Stretching the body	An eagle gliding to find food and feed its chicks

TABLE 8.2 Ideas rated low by domain experts in both usefulness and perceived causality

Working title	Target behavior	Analogous story
Scoop	Eating a meal steadily to the end	Using a paper scoop to play a goldfish catching game
Good Meal	Eating a takeout lunch to the end	Saving animals trapped in snow
Tooth Barista	Brushing teeth thoroughly	Preparing froth for a morning coffee
WagOut	Walking the dog	A chef scouting for truffles in the forest with their dog
Washie	Doing laundry at home	Using a machine to incubate eggs for chicks

can see the possibility of the target behavior and the story happening together. This can be achieved through remembering a similar scenario or retrieving a commonly known concept, wherein the behavioral acts and the story events might co-occur. In other words, one invokes or retrieves a knowledge frame from long-term memory that includes relations useful in combining the behavior and the analogous story in a nuanced way. Components, respectively, from two sides can be re-arranged in the invoked knowledge frame, forming a third story wherein a behavioral act seems to cause a story event. A high average score in perceived causality suggests that the raters could commonly invoke an external knowledge frame to reinterpret and make sense of the combined scenario shown in the middle panel. Later sections in this chapter will discuss the possible knowledge frames for some of these high-rated ideas.

On the other hand, perceived usefulness of the idea depends on individual experts' perspective. As the context of the question is about the behavior that the target user wants or needs to do, the interpretation of usefulness here can be associated with whether the mixed-reality experience can make the target user more likely to perform the behavior. In other words, a high average score in usefulness indicates that the raters generally perceived the idea to be persuasive or motivating.

Hence, the eight ideas consensually rated high in both perceived causality and usefulness seem to demonstrate expressive iteration in that blending the target behavior with an appropriate analogous story yield an imaginative third story, which gives new reasons for performing the behavior. The blend in each idea requires invocation of a knowledge frame and entails causal relations between the behavioral acts and the story outcomes, resulting in a strong sense of perceived causality. The story outcomes at different stages are tied with the corresponding behavioral acts, because the former is like an intrinsic part of the latter. Then, the third story unfolds along a direction in line with one's primary reasons (purposes or benefits) for performing the behavior. The story ending and one's vision are coherent and even resonant. When a person is absorbed in the story, they also become interested in the behavior. This might explain why a rater gives such ideas a high usefulness score.

Conversely, the five ideas consensually rated low in both perceived causality and usefulness imply that raters in general failed to recall or invoke knowledge frames that could explain the connections between the behavioral acts and the story outcomes. Without a knowledge frame, the components, respectively, from two sides are just put together without a context. In other words, the blend is not very sensible. The analogous story just develops on a trajectory away from one's original vision. In the coming sections, I will also discuss the lack of knowledge frames in some cases.

Other than the above highly rated or lowly rated groups, two ideas that were rated moderate in usefulness are worth mentioning here for comparative analysis. WaterBot is an idea that was rated moderate to high in perceived causality and moderate only in usefulness. It targets the running routine, which can be juxtaposed with the Mario's running story discussed in Chapter 4. Hatcher is another idea that was rated moderate in both perceived causality and usefulness. It targets the same behavior as Sitravel does, the other highly rated idea. Comparing the two ideas targeting the same behavior with different analogous stories would shed light on how they work differently.

In summary, out of the 35 ideas included in the expert evaluation study, some highly rated ideas, lowly rated ideas, and moderately rated ideas are worth further analyzing. They cover an array of lifestyle behaviors that can be integrated into daily routines. Reorganizing them in terms of the target behaviors, and merging them with the routine list summarized in Parts I and II, I consolidate a fairly extensive list of routines, separated into goal-directed and maintenance categories. Table 8.3 presents the list of goal-directed and maintenance routines, each of which is followed by potential analogous stories for comparison, and possible knowledge frames for sensible blends. In the following sections, I will discuss some of them regarding the suggested knowledge frames and the resonance, if any, between the analogous stories and the routines.

Goal-Directed Routines

Running

Two different analogous stories are considered for blending with the running routine. The Super Mario Bros. running story is discussed at length in Chapter 4. A person runs on the street. Mario runs on the platform in the game. For the two running events to co-occur, an external knowledge frame [running buddies] can be invoked together with the relation BUDDIES () that makes the person and Mario become running buddies, who practice together. Another imported relation SIDE BY SIDE () confines the person and Mario passing the same point at the same time. The primary reason for performing the running routine is generally to achieve or maintain physical fitness. To be specific, the person might want to prepare for an upcoming marathon. On the other side, Mario's story continues to unfold, saving

TABLE 8.3 A consolidated list of goal-directed and maintenance routines with corresponding analogous stories and suggested knowledge frames, if any

Goal-directed routine	Analogous story	Knowledge frame
Running	Super Mario Bros. running (Ch.4)	Running buddies
	An airplane flying and refueling intermittently (WaterBot)	?
Workouts	The Turnip	Urban farming
	The Fallen Tree	City trails
Taking medications	The Three Little Pigs	Neighborhood
Attending classes	The Flying Ship	Team project
	How Eleven Made Their Way in the League	Team project
Commuting to do piecework	Tsarevitch Ivan, the Firebird and the Gray Wolf	Mission titles, named trains
Trading used items	The Stonecutter	Chain reaction
	The Chef	Chain reaction
Dollar cost averaging	*Incomplete Open Cubes*	Blockchain technology
	Incomplete Asanoha	Blockchain technology

Maintenance routine	Analogous story	Knowledge frame
Drinking adequate water	Watering and growing the plant (Water Plant)	Humidity
Taking stairs	*36 Views of Mt. Fuji*	Observation tower
	Hiking to enjoy spectacular views (Staircase Hiking)	Observation tower
Having lunch	*100 Famous Views of Edo*	Public space
	Using a paper scoop to play a goldfish catching game (Scoop)	?
	Saving animals trapped in snow (Good Meal)	?
Focusing on desk work with breaks	Traveling on a train to visit attractions (Sitravel)	Office cars
	A hen hatching eggs in the nest (Hatcher)	?
Doing laundry at home	Traveling on the Tōkaidō	Accommodation
	Using a machine to incubate eggs for chicks (Washie)	?
Walking the dog	Counting days on nine-nines charts	Visual journal, calendar
	A chef scouting for truffles in the forest with their dog (WagOut)	?
Stretching	An eagle gliding to find food and feed its chicks (Stretching Wings)	Giant bird riding
Breathing exercise	Blowing beautiful bubbles (Mood Bubbles)	Mental imagery
Organizing the schoolbag	Feeding the monster pet appropriately (Feed Your Bag)	Familiars
Self-monitoring peak flow	Practicing the flute (Peak Flow Flute)	Music recording

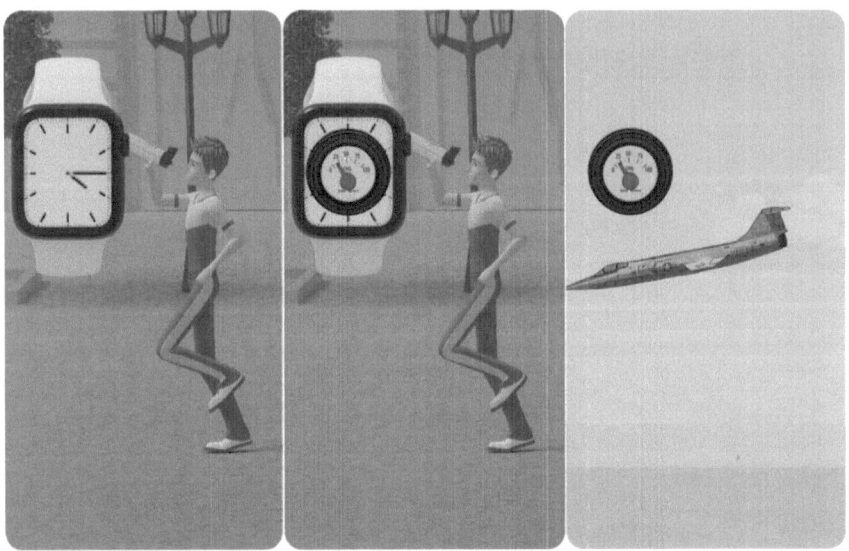

FIGURE 8.4   Screenshots of the three panels showing the running routine (left), the analogous story of a flying airplane (right), and the blended third story of seeing the water level and the oil level synchronized (middle).

the princess at last. Can this story ending be coherent with the person's vision, that is, physical fitness? As the person and Mario are buddies, they not only practice together but also help each other. When the person is physically fit, one is also able to help Mario save the princess. They are coherent. In WaterBot, a person runs on the street and intermittently stops by a dispenser to refill the water bottle. The running routine includes acts slightly different than the previous routine. The analogous story visualizes an airplane flying in the air and landing at an airport to refuel (see Figure 8.4). It is not difficult to structurally align the routine acts with the story events. Yet, on what occasion would the person's running be in synchronization with the airplane? Is it possible that the water level in the bottle is tied with the oil level in the tank of the airplane? It does not obviously recall a knowledge frame that makes the connection possible. Hence, experts generally rated this idea only moderate in perceived causality. In the analogous story, the airplane might fly and land at different airports to refuel. It might travel to many different airports until retiring in the end. Without an invoked knowledge frame for the blends, it is hard to relate the story ending to the person's vision. Flying to different airports and physical fitness seem unrelated. One might unconvincingly argue that the person stays physically fit to travel to different cities hosting marathons. Alternatively, though again far-fetched, the person could be a pilot of the airplane. But this requires more assumptions than most people can think of. The raters could not see the connection, and the perceived usefulness was rated moderate only. Table 8.4 lists out the considerations for the blends.

TABLE 8.4 Knowledge frame not found in the blends between the running routine and the story of a flying airplane, and misalignment between the routine person's vision and the story ending, as listed in the last row

Running	Selective projection and pattern completion	An airplane flying and refueling intermittently (WaterBot)
	[what knowledge frame?]: SYNC (water level, oil level)?	
RUN ON (person, street) →	RUN ON (person, street) FLY (airplane, air)	TAKE OFF (airplane, airport) ← FLY (airplane, air)
STOP BY (person, dispenser)	?	LAND (airplane, airport) REFILL (airplane, gas)
REFILL (person, bottle)		
…	…	…
Physical fitness	Physically fit to travel to different cities for marathons? Physically fit to be a pilot?	Airplane successively landing at different airports before retiring

Workouts

A workout routine is different from a running routine. The former is usually performed in a gym. Moreover, a workout routine involves varied specific bodily actions that engage different muscles, such as lifting, pulling, and pushing. It can be associated with different analogous stories, including "The Turnip" and "The Fallen Tree," which are discussed in Part II. Both stories involve repetitive acts of pulling or lifting. In "The Turnip," characters pull up a turnip in the farm. For that act to co-occur with workouts in the gym, the external knowledge frame [urban farming] can be invoked to conjure up the image of an urban farm near the gym. The person working out daily can see the grandfather pulling the turnip in the farm. One might even join the family to help. The causality between the workout act and how much the turnip has been pulled out can be easily perceived. The primary reason for workouts is typically for muscle building. Some people might even aim at signing up for bodybuilding contests. The story developments of "The Turnip" resonate with one's vision in doing workouts. Building muscle is a way to boost physical strength, which in turn can be instrumental in pulling gigantic root vegetables out of the soil. While a person working out might target a bodybuilding contest, the grandfather aims at farmers' contests for agricultural produce. Chapter 7 suggests "The Fallen Tree" as a new variant of "The Turnip," presenting a group of volunteers who lift a fallen tree that blocks a trail instead of the farmer family. There are hiking trails near urban areas in some cities (e.g., Hong Kong is well-known for her accessible trails). The knowledge frame [city trails] can be invoked to make sense of a scenario where a trail is near the gym. A person working out can see the trail from the gym, or pass by it on the way to the gym. Seeing volunteers work together to clear the trail, one might join to help. The causality between doing workouts and the trail condition can be imagined. Again, muscle strength is essential in manually clearing and maintaining the trail. The person's primary reason for workouts is coherent with the trail maintenance story. Table 8.5 summarizes the blending and convergence of the workout routine and "The Fallen Tree."

Taking Medications

The act of taking medications or supplements involves specific actions like taking pills out of dispensers or bottles, swallowing pills, or using a spoon for liquid medicine, which are not easily found and matched in other scenarios or stories. Chapter 5 suggests "The Three Little Pigs" for blending with the routine. The act of building houses in the analogous story is not the same as the routine act. But they share comparable purposes of building protection from danger. More importantly, the external knowledge frame [neighborhood] enables the imagination of a scenario where the pigs' houses are near the person's home. Due to the wolf's attacks, building new houses and taking medications can co-occur. The causality between the story events and the routine cycles can be perceived. This case shows that similarities between the routine and the analogous story might not lie in common sensory images (e.g., tactile or kinesthetic senses) or basic image schemas (e.g., proximity or containment relations), but more complex image schemas like causal relations.[13] The person needs to take medications because they have caught a cold, while the pigs need to build houses again because the old ones are not resilient enough.

The primary reason for taking medications or supplements is of course to achieve physical health. More specifically, one wants to strengthen the immune system. It resonates with the ending of "The Three Little Pigs." When the person is physically strong, they can help others defend themselves against any attacks (telling the pigs that the wolf is climbing over to the chimney).

Attending Classes

In Chapter 5, I suggest blending the routine of attending classes with the folktale "The Flying Ship" because they share similar repetitive patterns in terms of acquisition. In every class, the person acquires knowledge or skills on one topic. After attending all classes, the person is fully equipped. In the analogous story, the simpleton meets and invites one person with a special skill on board at a time. Before arriving at the castle, he has built a full team. The external knowledge frame [team project] allows co-occurrence of attending classes to learn and inviting strangers on board the ship. In every class, the person imagines meeting a talented classmate in a team project and learning a skill from their peer. The causality between attending classes and building a team on board can be imagined.

TABLE 8.5 The blends between the workout routine and "The Fallen Tree," plus the convergence between the routine person's vision and the development of the story, as listed in the last row

Workouts	Selective projection and pattern completion	The Fallen Tree
	[city trails]: NEAR (trail, gym) SEE (person, tree) PULL (tree, volunteer1)	← PULL (tree, volunteer1)
PULL (person, → Day1)	PULL (person, Day1) UP (tree, 15%) [city trails]: NEAR (trail, gym) SEE (person, tree) PULL (tree, volunteer1, volunteer2)	← UP (tree, 15%) ASK (volunteer1, volunteer2) ← PULL (tree, volunteer1, volunteer2)
PULL (person, → Day2)	PULL (person, Day2) UP (tree, 30%) [city trails]: NEAR (trail, gym) SEE (person, tree) PULL (turnip, volunteer1, volunteer2, volunteer3)	← UP (tree, 30%) ASK (volunteer2, volunteer3) ← PULL (tree, volunteer1, volunteer2, volunteer3)
PULL (person, → Day3)	PULL (person, Day3) UP (tree, 45%)	← UP (tree, 45%)
…	…	…
Building muscles	Muscular enough to help clear any fallen trees on the trail	Clearing and maintaining the trail

The primary reason behind attending classes is generally for academic or personal development. People might have more specific reasons for enrolling different courses, for example, aiming to learn a foreign language for study or life abroad. Similar reasons behind attending classes can be aligned with the ending of "The Flying Ship." The simpleton gets help from his team to go through all the challenges and returns home with experience and rewards. After completing the course, the person also finds the experience fruitful and becomes confident to take future challenges. In Chapter 7, I suggest a new variant of "The Flying Ship." "How Eleven Made Their Way in the League" uses the plot with similar repetitive patterns, but is varied in the characters and settings. A soccer player-turned manager invites different talented players to join his new soccer team. For blending with the routine, the knowledge frame [team project] applies again and provides a scenario for both attending classes and building a team with strong players to co-occur. The causality can be perceived too. The soccer team story can end with the team entering the community league tournament and winning their way to the championship.

TABLE 8.6 The blends between the routine of attending classes and the new variant "How Eleven Made Their Way in the League," plus the convergence between the routine person's vision and the development of the story, as listed in the last row

Attending classes	Selective projection and pattern completion	How Eleven Made Their Way in the League
ATTEND (person, class, Week1) →	ATTEND (person, class, Week1)	
	TALENT (goalkeeper, skill1)	← TALENT (goalkeeper, can catch at lightning speed)
	[team project]: PEER (person, goalkeeper)	INVITE (manager, goalkeeper)
	LEARN FROM (person, goalkeeper)	ON BOARD (retired, goalkeeper)
LEARN (person, skill1) →	LEARN (person, skill1)	
ATTEND (person, class, Week2) →	ATTEND (person, class, Week2)	
	TALENT (striker, skill2)	← TALENT (striker, can shoot at any angle)
	[team project]: PEER (person, striker)	INVITE (manager, striker)
	LEARN FROM (person, striker)	ON BOARD (retired, goalkeeper, striker)
LEARN (person, skill2) →	LEARN (person, skill2)	
ATTEND (person, class, Week3) →	ATTEND (person, class, Week3)	
	TALENT (left wing, skill3)	← TALENT (left wing, can run faster than a cheetah)
	[team project]: PEER (person, left wing)	INVITE (manager, left wing)
	LEARN FROM (person, left wing)	ON BOARD (retired, goalkeeper, striker, left wing)
LEARN (person, skill3) →	LEARN (person, skill3)	
…	…	…
Academic development	With skills and experience to champion in a domain	The full team heading to the community league

This perfectly matches the vision of the person signing up for a course for academic development. After completing the course, the person is equipped with skills and experience to outcompete others in a specific domain (see Table 8.6).

Commuting to Do Piecework

Daily commute can be a maintenance routine, but commuting to do piecework, like food delivery or on-site handyman jobs, bears a clearer goal – getting over with it and moving on to the next one. Chapter 5 suggests blending this kind of commute with the chain tale "Tsarevitch Ivan, the Firebird and the Gray Wolf." They share similarities in riding a vehicle (i.e., kinesthetic) and repetitive patterns in finishing given tasks. The person might ride a train to a place for a job, receive another job order, and travel again. Ivan rides the Gray Wolf to accomplish a "stealing" mission, gets caught, and is given another mission again. For the real-life routine acts and the fairytale events to co-occur, two knowledge frames are invoked. The frame [named trains] allows one to see the train as the magic creature. The frame [mission titles] allows the person to name the jobs they get the same as Ivan's stealing targets. Commuting to do piecework every day is mapped with Ivan accomplishing a given mission. This renders the causal relation between the routine acts and the fairytale events. In the end of the chain tale, Ivan, with the help of the Gray Wolf, acquires fruitful rewards and returns home. This perfectly aligns with the primary reasons for the routine.

Trading Used Items

Some lifestyle behaviors involving acts of exchanges can also be practiced as routines, such as trading used items for new ones, or recycling used containers in exchange of discount coupons or other incentives. When practicing these behaviors, a person might have a long-term vision or belief such as economic or environmental sustainability, but still rely on some favors or immediate rewards to trigger their actions. In Chapter 5, I suggest blending this kind of routine with another chain tale "The Stonecutter." They share similar repetitive patterns in exchanging current possessions for something new. In reality, the person trades a used item for a new one. In the fairytale, the stonecutter wishes to be another powerful being instead of his current self. The separation of reality and fairytales renders blending them non-trivial. One possible knowledge frame can be [chain reaction], which entails propagation of exchange events from one side of a huge ecosystem to another side. Whenever the person trades a used item in reality, this triggers the chain reaction and leads to the stonecutter making a new wish. This causal relation, though distant, can be imagined when the exchange nature is highlighted.

In the ending of "The Stonecutter," the protagonist goes through the transformations and realizes the unique capability and value of each being. He gains experience and learns a lesson. This ending matches the primary reasons for such routines as trading items or recycling used containers. Through practicing similar routines, the person experiences the use of different goods and realizes the virtue of recycling, resonating with the stonecutter's enlightenment. In Chapter 7, I suggest a new variant of "The Stonecutter," which I call "The Chef." It alludes to the career development of a real-world chef, André Chiang, hailed from Taiwan. He started helping in his mother's small restaurant in Japan when he was a teen, and subsequently worked in different fine-dining restaurants in France and other places, before moving back to Taiwan to open restaurants. Whenever he got employed by another restaurant, he had to quit his chef job in the previous restaurant. His story is also a sequence of repetitive exchanges of the old position for the new one. To blend the chef story with the routine of trading used items, the same knowledge frame [chain reaction] still applies. Whenever the person trades a used item for a new good, one might imagine that the act of trading triggers a chain of exchanges, probably including someone bringing a designer handbag to a fine-dining establishment which leads to the chef quitting and joining another Michelin-starred restaurant.

After acquiring substantial experiences in different restaurants across the globe, the chef recently moved back to his home country and focused on original creations using authentic local ingredients. His recent focus on authentic local ingredients and the person's vision in trading used items for the purposes of sustainability are coherent (see Table 8.7).

Dollar Cost Averaging

Dollar cost averaging (DCA) suggests regularly using the same amount of money to buy stocks in installments. In a long run, the average cost of purchasing the shares is optimal. Sticking to this routine requires self-discipline and perseverance. Drawing on Sol LeWitt's serial art project, *Incomplete Open Cubes*, I suggest in Chapter 6 that the artist's systematic process of cataloging all variations of incomplete open cubes can be considered as an analogous story for DCA. The cataloging process requires successive inputs of mental efforts to think of exhaustively the possible configurations of an open cube, which demands one's determination and persistence to enumerate, just like the long-term investing routine. The cataloging endeavor and the DCA routine share similarities in repetitive investments, respectively, mental efforts and capital. The knowledge frame [blockchain technology] is intriguingly relevant to their co-occurrence in an imagined context. To append a new transaction to a blockchain, "proof-of-work" is required to validate the authenticity of the act of purchasing. Rearranging the edges and finding a new variation of incomplete open cubes can symbolize the proof of invested efforts. The causal relation between the purchase and the newly found variation is clear to someone who knows about the basic principle of blockchain technology.

The primary reason for DCA is obviously the growth of wealth. The result of the artist's serial project is a substantial catalog of 122 variations. The growth of wealth is echoed by the growth of the catalog of found variations. In Chapter 7, I suggest a new and culturally charged variant of the originally pure serial art project. Instead of the open cubes, I suggest using a traditional Japanese pattern, "asanoha." The pattern is based on a regular, dissected hexagon, with three short segments added to each of its equilateral triangles, visually resembling hemp leaves and symbolizing growth and success. The cataloging process of the original serial art project can be applied to the asanoha pattern, starting from a regular, dissected hexagon, adding one segment at a time, and the person comes up with all variations of incomplete

TABLE 8.7 The blends between the trading routine and the new variant "The Chef," plus the convergence between the routine person's vision and the development of the story, as listed in the last row

Trading used items	Selective projection and pattern completion	The Chef
OWN (person, item1) →	OWN (person, item1)	
TRADE-IN (person, item1, new item2) →	TRADE-IN (person, item1, new item2)	
	DREAM (Japan, France) [chain reaction]: PROPROGATE (EXCHANGE (item1, new item2), EXCHANGE (Japan, France))	← DREAM (Japan, France)
	MOVE (Japan, France)	← MOVE (Japan, France)
OWN (person, item2) →	OWN (person, item2)	
TRADE-IN (person, item2, new item3) →	TRADE-IN (person, item2, new item3)	
	DREAM (France, Thailand) [chain reaction]: PROPROGATE (EXCHANGE (item2, new item3), EXCHANGE (France, Thailand))	← DREAM (France, Thailand)
	MOVE (France, Thailand)	← MOVE (France, Thailand)
OWN (person, item3) →	OWN (person, item3)	
TRADE-IN (person, item3, new item4) →	TRADE-IN (person, item3, new item4)	
	DREAM (Thailand, China) [chain reaction]: PROPROGATE (EXCHANGE (item3, new item4), EXCHANGE (Thailand, China))	← DREAM (Thailand, China)
	MOVE (Thailand, China)	← MOVE (Thailand, China)
...	...	...
Economic sustainability Environmental sustainability	Gaining experience of enjoying different goods and finally realizing the value of authenticity	Gaining experience and back to home country for original creations and authentic ingredients

asanoha. The knowledge frame [blockchain technology] applies the same way. Every variation is the proof-of-work for each authentic transaction. The perceived causality is clear.

The catalog of asanoha patterns can be seen as a kumiko, traditional Japanese woodwork panel. The panel, if virtually displayed on a large screen or projected on

TABLE 8.8 The blends between the DCA routine and the new variant project Incomplete Asanoha, plus the convergence between the routine person's vision and the results of the variant project, as listed in the last row

Dollar cost averaging	*Selective projection and pattern completion*	*Cataloging variations of Incomplete Open Asanoha*
PURCHASE (person, stocks)	→ PURCHASE (person, stocks)	
TRANSACTION ($100, stocks)	→ TRANSACTION ($100, stocks) [blockchain technology]:	REARRANGE (master, 1 segment)
	FIND (person, 1-segment asanoha)	← FIND (master, 1-segment asanoha)
	LOG (person, 1-segment asanoha)	← LOG (master, 1-segment asanoha)
	PoW (1-segment asanoha) APPEND TO LEDGER (TRANSACTION ($100, stocks), PoW (1-segment asanoha))	
...	...	...
Economic sustainability Wealth growth	Wealth gradually grows with the hemp field kumiko	A kumiko of a huge hemp field built

the wall, looks like a prosperously grown hemp field, which elegantly symbolizes the person's vision in DCA, that is, wealth growth (Table 8.8).

Maintenance Routines

Taking the Stairs

Regularly taking the stairs instead of an elevator in daily life not only increases one's physical activity level and benefits physical health, but also benefits the environment in a long run. With more people practicing stair climbing, the demands of elevators or escalators in public space would dwindle, leading to less power consumption. Stair climbing is promoted in many urban areas of today. A person might start a daily routine of climbing stairs to their office or back home. By applying expressive iteration, one can render the routine more satisfying and engaging. In Chapter 5, I suggest blending the routine with the travel narrative of the ukiyo-e print series *36 Views of Mount Fuji*. When the person climbs the stairs to a particular floor, the altitude of the floor corresponds to a virtual view of Mount Fuji at a similar level selected from the series. Stair climbing and the ukiyo-e travel narrative share similarities in the act of ascending a structure or an elevated Earth surface (i.e., kinesthetic). The external knowledge frame [observation tower] enables an imaginative situation wherein climbing stairs and reaching a viewpoint can co-occur. The imported relation DECK (floor, altitude) maps different floors with different altitudes, determining which viewpoints and views from the wood-block print series are most appropriate. Coincidently, an idea collected from the workshop series, Staircase Hiking (see Figure 8.3), proposes similar blends with a simple analogous story of hiking in a mountain and reaching the top to enjoy the spectacular view. The original idea only presents generic locations and views, which can be complemented by the ukiyo-e print series. In the expert evaluation study, Staircase Hiking was presented with two episodes, one showing a generic mountain and the other resembling Mount Fuji. The perceived causality was rated high (3.75), indicating that the correlation of climbing stairs and enjoying sweeping views from somewhere high up is obvious.

The raters also perceived Staircase Hiking to be useful, as reflected in the average score (3.5). The ending of the travel narrative is traveling around the region, visiting all recommended locations, and capturing all spectacular views. This is in line with the primary reasons for taking stairs, including physical health and environmental concerns. The person climbing stairs can stay physically healthy to travel around, hike, and appreciate the spectacular views in the sustainable environment.

Having Lunch on Workdays

Many working urbanites do not have a long lunch break. Many resort to buying a quick and convenient takeout lunch, or having a packed meal. Apart from eating the takeout or packed lunch at the workplace, which might not be always appropriate, one could also find a quiet corner in some public space and enjoy the meal. This can be a common maintenance routine on workdays. In Chapter 5, I suggest blending similar routines with the travel narrative of *100 Famous Views of Edo*. Having a takeout or packed lunch in a public space is similar to the ukiyo-e travel narrative because both involve finding a favorite spot (i.e., kinesthetic). The external knowledge frame [public space] enables the person to imagine the co-occurrence of finding a seat to eat a takeout lunch and finding a spot to enjoy great views. Having found a seat to enjoy the meal, one also enjoys the views of the public space. The causality between the two sides seems obvious and can be perceived.

Having takeout lunch on workdays is a way to cope with the short lunch break. One might also want to save money because of the high living cost in the city. Meanwhile, healthy and mindful eating is also important. While enjoying a takeout lunch in a public space, one can enjoy the surroundings and also virtually see a famous view selected from the woodblock print series. This experience is in line with the primary reasons for eating. A number of ideas collected from the workshops serve to address eating habits, like slow and mindful eating, or finishing all food that one orders. Scoop is one of the ideas that suggests blending the routine of eating a packed or takeout lunch with a story of using a paper scooper to play the traditional Japanese goldfish catching game (Figure 8.5). In the game, a player

FIGURE 8.5   Screenshots of the three panels showing the routine of having a takeout lunch steadily (left), the analogous story of using a paper scoop to catch goldfish in the pool (right), and the blended third story of catching goldfish while eating a takeout meal (middle).

needs to use the paper scooper to move a goldfish from the pond to their bucket. The trick is to move gently and steadily. This required tricky act is similar to the expected steady movements in mindful eating (i.e., kinesthetic). The paper scooper slightly resembles a spoon used for eating packed or takeout lunch (i.e., visual appearance). However, it is not easy to bring in any knowledge frame that can enable a situation wherein having lunch and catching goldfish might take place simultaneously. In the expert evaluation study, raters were likely to have encountered such a difficulty and could not imagine how they might be compatible. Hence, the average score in perceived causality is low (2.38). The goal of the game is to catch as many goldfish as possible, which is also irrelevant to one's vision in mindful eating. It could be argued that when the person is mindful, he or she can be more concentrated on catching goldfish. But the results of the game are not related to eating, as the goldfish is definitely not food (see Table 8.9).

Another idea from the workshop collection that also addresses the topic of eating is Good Meal. When buying a takeout lunch, a person can order the portion size according to their appetite to avoid food waste. The idea suggests blending the routine of ordering and finishing the lunch with an analogous story of saving animals trapped in snow (Figure 8.6). In the story, a rescuer cautiously shovels snow to uncover the animal trapped underneath. This story and the eating routine share similarities in several aspects. Both snow and rice are white in color (i.e., visual).

TABLE 8.9 Knowledge frame not found in the blends between the routine of having takeout lunch and the analogous story of using a paper scoop to play a goldfish catching game, and misalignment between the routine person's vision and the goal of the game, as listed in the last row

Having lunch	Selective projection and pattern completion	Using a paper scoop to play a goldfish-catching game (Scoop)
	[what knowledge frame?]:	
MOVE (spoon, food)	→ MOVE (spoon, food) MOVE (spoon, goldfish)?	← MOVE (scoop, goldfish)
CAUSE (STEADY (spoon), DIGEST (food))	→ CAUSE (STEADY (spoon), GAIN (goldfish) DIGEST (food))?	← CAUSE (STEADY (scoop), GAIN (goldfish))
…	…	…
Eating a meal at a steady pace Mindful eating	Eating steadily and catching more goldfish?	Catching maximum number of goldfish

FIGURE 8.6 Screenshots of the three panels showing the routine of having a takeout lunch to the end (left), the analogous story of saving animals trapped in snow (right), and the blended third story of saving trapped animals after finishing a takeout meal (middle).

TABLE 8.10 Knowledge frame not found in the blends between the routine of having takeout lunch and the analogous story of saving animals trapped in snow, and misalignment between the routine person's vision and the developments of the story, as listed in the last row

Having lunch	Selective projection and pattern completion	Saving animals trapped in snow (Good Meal)
ON (rice, dish) →	ON (rice, dish)	
WHITE (rice)	[what knowledge frame?]:	WHITE (snow)
	ON (rice, animal)?	← ON (snow, animal)
MOVE (spoon, rice) →	MOVE (spoon, rice)	MOVE (shovel, snow)
EMPTY (dish) →	EMPTY (dish)	
	UNCOVER (animal)?	← UNCOVER (animal)
...	...	...
Eating a meal steadily to the end	Finishing a meal and saving animals?	Saving different animals
Avoiding food waste		

The act of shoveling snow is like the act of spooning rice. They hold comparable spatial relations like rice "on" the dish and snow "on" the animal. Yet, the dish and the animal differ in scale. It is very difficult to think up any scenario wherein an animal is found on a dish after all rice is spooned. In the evaluation study, raters might not be able to imagine how and why a "small" animal is covered in the rice. They could not make sense of the connection between finishing the rice and finding an animal. The perceived causality was thus consensually rated low (2.06). In the rescue story, different animals are found and saved. The person ordering and finishing a meal should have a vision of avoiding food waste. Saving different animals seems unrelated to avoiding food waste. Could one argue that finishing a meal can give the rescuer energy to save animals? (See Table 8.10) It could be, but raters in the evaluation study generally could not connect the two sides. The average usefulness score was low (2.25).

Doing the Laundry at Home

Doing the laundry is definitely a common maintenance routine done by many. Although washing machines are now commonly available to those living in urban or suburb areas, the routine still includes a sequence of tedious steps, like sorting dirty laundry according to their colors and fabrics, operating the machine, drying, folding, and putting away the clean laundry. In Chapters 5 and 6, I suggest blending the laundry routine with the ukiyo-e pictorial narrative about traveling on the main roads connecting Edo and Kyoto, including the Tōkaidō and the Nakasendō (aka the Kisokaidō). Doing the laundry at home and traveling on the Tōkaidō or the like may seem to be dissimilar. In fact, they share a subtle yet inevitable repetitive act—washing clothes. During a long trip, a traveler cannot bring along too much luggage and needs to intermittently wash clothes in the guesthouse or hotel. The knowledge frame [accommodation] can enable the person performing the routine to imagine a scenario where they have to wash clothes on a trip, and the washing machine in the guesthouse might look like the one they have at home. The causal relation between having the laundry done and continuing the journey is sensible. The primary reason behind doing the laundry is to maintain household and personal hygiene. This vision focusing on home can be intriguingly linked to travel out of home. The laundry routine can develop a person's good hygiene habit and self-management ability, which are essential for a traveler on a long trip. The laundry routine can thus be imagined as training in preparation for an upcoming travel, where a traveler needs to keep their clothes clean and fresh. In other words, the routine and the travel narrative are coherent. An interesting idea from the design workshop dealing with the laundry routine is Washie. It suggests blending the routine with an analogous story of using a machine to incubate eggs for different chicks (Figure 8.7). The routine and the analogous story mainly share similarities in the act of putting something into a machine (i.e., kinesthetic) and the resulting spatial relation between the object and the machine, such as the laundry "in" the washing machine compared with an egg "in" the incubator. After washing, the dirty laundry becomes clean. Meanwhile, a chick is hatched in the incubator after some time. While the chicks might look adorable, the raters in the evaluation study seemed to find difficulty in connecting the routine act and the virtual chicks, resulting in the lowest average perceived causality score (1.88) among the selected ideas. In fact, it is difficult to imagine a possible scenario consisting of a machine that can turn dirty laundry into clean clothes, as well as an egg into a chick. Although the washing machine and the incubator both use heat in their processes (i.e., a similarity in the tactile sense), the overall mechanisms are so different that a common knowledge frame cannot be easily associated.

In the hatching story, incubating different eggs results in different adorable chicks. A collection of chicks motivates the behavior. Yet, this story ending is hardly related to the primary reason for doing laundry. One might argue that the routine develops good household hygiene and management ability, which are essential for managing a laboratory where chicks are hatched. This is a possible alignment, but the experience of laboratory management is less common and more difficult to be imagined, compared with the experience of keeping clothes clean during travel (see Table 8.11). Anyhow, the raters could not see the link and the average usefulness score was also the lowest (2.0).

Focusing on Desk Work with Breaks

In recent years, a number of studies have found that working in a sitting position for prolonged periods of time is associated with health issues such as metabolic syndrome (including diabetes) and cardiovascular disease. A sedentary lifestyle would cause a higher risk of mortality.[14] Many wearable devices of today include

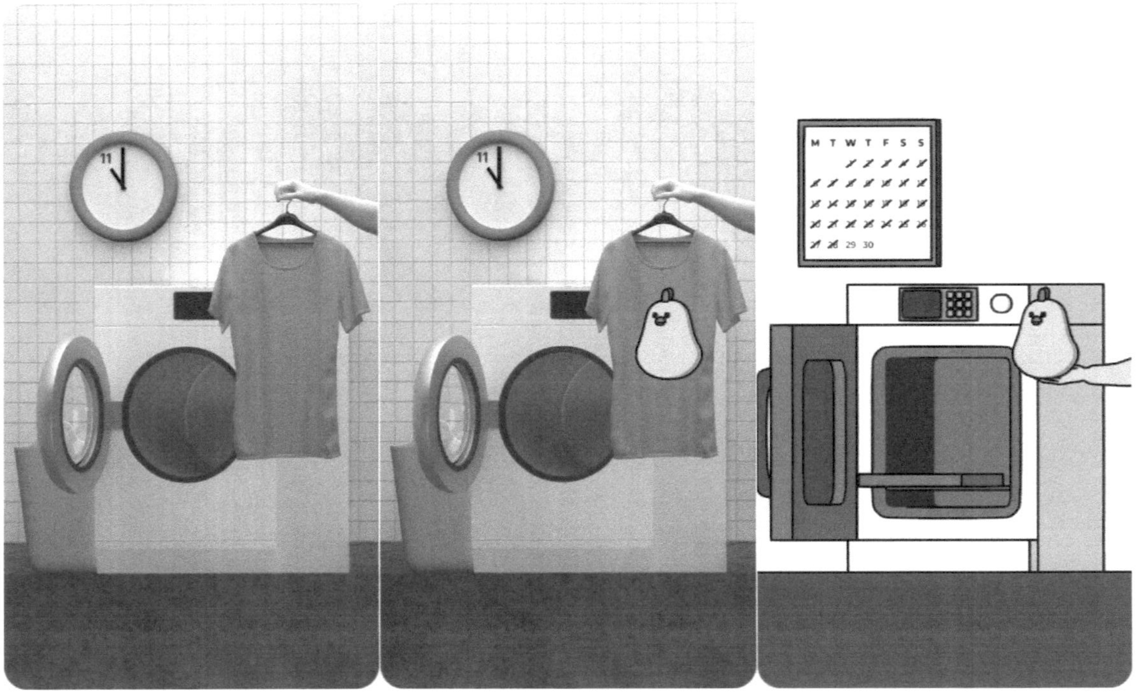

FIGURE 8.7 Screenshots of the three panels showing the laundry routine (left), the analogous story of using a machine to incubate eggs for chicks (right), and the blended third story of getting a chick hatched out after finishing laundry (middle).

TABLE 8.11 Knowledge frame not found in the blends between the laundry routine and the analogous story of using a machine to incubate eggs for chicks, and misalignment between the routine person's vision and the developments of the story, as listed in the last row

Doing laundry	Selective projection and pattern completion	Using a machine to incubate eggs for chicks (Washie)
	WARM (washing machine, incubator) [what knowledge frame?]:	
PUT INTO (dirty clothes, washing machine) →	PUT INTO (dirty clothes, washing machine)	
WASH (dirty clothes, clean clothes)	PUT INTO (egg, washing machine)? ←	PUT INTO (egg, incubator)
TAKE OUT (clean clothes, washing machine)	HATCH (egg, chick)? ← TAKE OUT (chick, washing machine)?	HATCH (egg, chick) TAKE OUT (chick, incubator)
…	…	…
Household hygiene and management	Good household hygiene for hatching chicks?	Hatching different chicks

a feature that tracks the user's sitting time and vibrates to remind them to stand up or take a break. A simple notification might work initially but users easily turn insensitive and ignore it after some time. Instead of labeling prolonged sitting a bad habit and sending regular reminders to prevent it, it is better to help users develop a routine of taking regular breaks after working or reading in a sitting position for some time. A few ideas from the workshop collection take this approach. Sitravel proposes blending the routine with an analogous story of traveling on a train to visit attractions along the line (Figure 8.8). When a person sits to work or read for an hour, they should stand up and take a short break, like getting a glass of water or stretching their arms, after which one can continue to work. When someone sits on a train and intends to visit different attractions near different stations along the line, he or she would stand up and get off the train when it arrives at a target station. One can take a picture of the attraction, and then get back on the train to continue the journey. The desk work with breaks routine and the travel story share similarities in the sitting position (i.e., kinesthetic), as well as the repetitive patterns of standing up (assuming the traveling time between stations to be about one hour or so). Yet, one typically works in a fixed location like an office, studio, or home, rather than on a moving train. Due to wide accessibility to the Internet and lightweight powerful notebooks today, as well as in part due to the COVID-19 pandemic, there have been changing work styles. One can work in the café, in the beach, or on a train. East Japan Railway Company has launched so-called "office cars" for certain shinkansen bullet train lines that connect Tokyo with the country's northern and central

FIGURE 8.8 Screenshots of the three panels showing the routine of focusing on desk work with breaks (left), the analogous story of traveling on a train to visit attractions (right), and the blended third story of sitting to work while traveling on a train's office car (middle).

regions. In an office car, a passenger may sit to work with distraction-blocking and noise-canceling gear in order to focus on work.[15] With the knowledge frame [office cars] invoked, one can imagine a scenario wherein sitting to work and traveling on a train actually co-occur. Having worked for an hour, one arrives at a station. He or she stands up and virtually sees an attraction on a wall-mounted digital picture frame simulating a car window. The causal relation between standing up and seeing an attraction is obviously sensible. The perceived causality was thus rated high (3.81) by the expert participants.

The original idea of Sitravel only touches on generic locations and attractions. The train travel story can be intriguingly integrated with the railway line of today connecting Tokyo and Kyoto roughly along the Edo-period Tōkaidō. The attraction near every train station can show the corresponding ukiyo-e print of each Edo-period station. Hence, one can see one print on the virtual car window when they stand up once every hour. This experience resonates with the person's original vision of taking breaks. The primary reason for working with regular breaks is to avoid sitting for too long, which is associated with negative health outcomes. Meanwhile, taking breaks also avoids burnout and refresh one's mind, bringing about mind stimulation. Seeing different road views on the Tōkaidō at every break is an enjoyable visual stimulus. Table 8.12 summarizes the blends.

Hatcher is another idea also addressing the routine of desk work with regular breaks. It suggests blending the routine with an analogous story about a hen hatching eggs in the nest (Figure 8.9). While a person should stand up after a bout of continuous sitting, the hen gives warmth to the eggs and has to stand up to give way whenever a chick hatches out. This analogous story, different from others, is an animal tale that expects the routine person taking the perspective of the animal. The routine and the hatching story share similarities in the sitting position (i.e., kinesthetic) and the tactile sense of warmth. Yet, the task chair and the nest look very different in both form and material (i.e., visual appearance). The repetitive patterns of standing up are also varied in the cyclical period. The person is recommended to stand up every hour, but a chick might hatch out intermittently. The animal tale together with the disparities renders finding a knowledge frame for blending very difficult. One might imagine a bizarre scenario where the person helps the hen and puts the eggs on the chair under their thighs. When a chick is about to hatch out, the person needs to stand up to free it. Yet, the expected weight of a human body makes one worry about the possibility of crushing the eggs rather than hatching them. It looks like raters also having similar doubt on whether sitting can cause hatching. Hence, the average score in perceived causality is only moderate (3.0).

The hatching story unfolds and may result in a collection of healthy and happy chicks. In the routine of desk work with regular breaks, one might envision prevention of negative health impacts from sitting for too long as well as mind stimulation. It is hard to see how a collection of chicks might contribute to this vision. Hence, the raters commonly judged the usefulness as only moderate too (3.0). Table 8.13 summarizes the issues in the blends.

TABLE 8.12 The blends between the routine of focusing on desk work with breaks and the analogous story of traveling on a train to visit attractions (e.g., on the Tōkaidō), plus the convergence between the routine person's vision and the developments of the travel narrative, as listed in the last row

Focusing on desk work (or reading) with breaks	Selective projection and pattern completion	Traveling on a train (e.g., on the Tōkaidō) to visit attractions (Sitravel)
	[office cars]	
SIT (person, chair, one hour) →	SIT (person, chair, one hour) ←	SIT (person, seat, one hour)
	ARRIVE (Fujisawa) ←	ARRIVE (Fujisawa)
STAND UP (person) →	STAND UP (person)	STAND UP (person)
	APPEAR (stream, bridge, travelers, torii)	GET OFF (person, train)
		VIEWPOINT (street-level view, No. 6)
		APPEAR (stream, bridge, travelers, torii)
	[office cars]	
SIT (person, chair, one hour) →	SIT (person, chair, one hour) ←	SIT (person, seat, one hour)
	ARRIVE (Kanbara) ←	ARRIVE (Kanbara)
STAND UP (person) →	STAND UP (person)	STAND UP (person)
	APPEAR (snow, travelers, huts)	GET OFF (person, train)
		VIEWPOINT (street-level view, No. 15)
		APPEAR (snow, travelers, huts)
...	...	...
Avoiding prolonged sitting	Taking breaks and visiting each station for mind stimulation	Visiting all stations and arriving at the destination
Mind stimulation		

Walking the Dog

Walking the dog is a unique maintenance routine. A person has to walk the dog every day, because dogs, particularly medium- to large-sized ones, need to walk outdoor to patrol their territories, socially interact with other dogs, and most importantly urinate and defecate. This is not an easy task if performed on a daily basis. Meanwhile, most dog-owners would like to walk the dog to build owner–dog relationship, as well as to boost the health of both themselves and their dogs. Yet, one might skip or procrastinate due to overwhelming schedules or other temptations. Applying expressive iteration to the routine would assist dog-owners in meeting the pets' wants and needs. In Chapter 6, I suggest that the routine can be blended with the daily logging narrative of using the traditional Chinese winter calendar charts, the nine-nines charts. When a person walks their dog, they see and remember the scenery on the way, such as the winding paths or colorful flowers in the

FIGURE 8.9 Screenshots of the three panels showing the routine of focusing on desk work with breaks (left), the analogous story about a hen hatching eggs in the nest (right), and the blended third story of sitting to work while helping to hatch eggs (middle).

TABLE 8.13 Knowledge frame not found in the blends between the routine of focusing on desk work with breaks and the analogous story about a hen hatching eggs in the nest, and misalignment between the routine person's vision and the developments of the story, as listed in the last row

Focusing on desk work (or reading) with breaks	Selective projection and pattern completion		A hen hatching eggs in the nest (Hatcher)
SIT (person, chair, one hour) →	SIT (person, chair, one hour) [what knowledge frame?]: SAME (chair, nest)?		SIT (hen, nest, one day)
	HATCH (person, egg)?	←	HATCH (hen, egg)
STAND UP (person) →	STAND UP (person)		STAND UP (hen)
	OUT (chick, egg)	←	OUT (chick, egg)
…	…		…
Avoiding prolonged sitting Mind stimulation	Chicks hatching out during breaks, stimulating thoughts?		All chicks hatching out and healthy

park, graffiti or window displays in the street, or sunset along a promenade. On the other hand, a person using a nine-nines chart to count the winter days writes one stroke each day and makes a note of the weather that day. After the first nine days have lapsed, they see a Chinese character (春) meaning "spring," which consists of pictographs resembling grass and the sun. The routine of walking the dog and the narrative of counting days on a chart share a similarity in the repetitive patterns of experiencing time passed. The knowledge frame [visual journal] provides reference for an imaginary scenario wherein perceiving the surroundings while walking the dog and recording the observations back home can take place. The perceived causality between the routine acts and the results on the chart is imaginable. When the winter is over, the person using the nine-nines chart can see the nine Chinese characters forming a poem with a poignant yet hopeful message. This is not only a caption for the pictures on the visual journal but also a remark for passing the bitter winter and anticipating the prosperous spring. The message perfectly aligns with one's vision in walking the dog, celebrating the fact that the owner and their pet have walked together and survived winter, staying healthy, maintaining their close relationship, and continuing to walk into the spring season with joy. Blending the routine with this narrative is perceived to be useful. An idea in the collection, WagOut, suggests blending the routine of walking the dog with an analogous story about a chef foraging truffles in a forest with their dog. While a dog-owner might walk the dog in urban areas, on the streets, or in a park, the chef walks in a forest with their dog to search for a precious ingredient for the chef's menu (Figure 8.10). The two sides share a similarity in the act of walking (i.e., kinesthetic) with a companion (i.e., visual and tactile). Yet, the settings, including other objects and environments, all look very different. It is not easy to imagine a scenario where both

FIGURE 8.10   Screenshots of the three panels showing the routine of walking the dog (left), the analogous story about a chef scouting for truffles in the forest with their dog (right), and the blended third story of walking the dog and foraging truffles on the street (middle).

TABLE 8.14 Knowledge frame not found in the blends between the routine of walking the dog and the analogous story about a chef scouting with the dog for truffles, and misalignment between the routine person's vision and the developments of the scouting story, as listed in the last row

Walking the dog	Selective projection and pattern completion	A chef scouting the forest with the dog for truffles (WagOut)
WALK (person, dog, street) →	WALK (person, dog, street) [what knowledge frame?]	SCOUT (chef, dog, forest)
NOSE (dog, something, postbox) →	NOSE (dog, something, postbox)	
PHUB (person, phone)	FOUND (dog, truffle, postbox)? ←	FOUND (dog, truffle, log)
	PICK UP (person, truffle)? ←	PICK UP (chef, truffle)
...	...	...
Physical health of both	Both the human and the dog being healthy and good partners to scout?	Collecting precious truffle to create the highly praised chef's menu
Human-dog relationship		

activities can happen simultaneously. Would the dog-owner become the chef? If so, the two activities would take place one after the other. Moreover, the key object, the truffle, has no equivalents in the street. If the virtual truffle just appears in the street without a physical counterpart, the user experience seems like a typical mobile AR game, such as Pokémon GO. The perceived causality between the physical act of walking and the virtual reward (i.e., truffle) lacks a knowledge frame as reference. The average score by the experts in the study was low (2.13).

In the chef story, the chef was able to find different kinds of truffles for the chef's menu on different days. The chef impressed customers as well as critics, earning acclaims and recognition. This can be a rags-to-riches plot, but it can hardly be associated with the human–dog relationship or their physical health. One might argue that the human and the dog need to develop good health and partnership in order to effectively forage truffles together. Yet, this connection is not highlighted in the analogous story. This might explain its low average score in usefulness (2.13) rated by the experts. Table 8.14 summarizes the issues in the blends.

Organizing the Schoolbag

Some routines are set to meet the needs of specific target groups. School kids are encouraged to organize their bags every night prior to a school day. Making it a routine has several benefits, not only saving time in the morning and reducing the risk of forgetting certain items but also helping a child develop their organization skills. Apart from reminders from parents or teachers, kids might need a nudge. Feed Your Bag is an idea from the workshop collection that attempts to motivate kids to tidy up their schoolbags. It suggests blending the routine with an analogous fantasy story about feeding a monster pet (Figure 8.11). The monster pet needs to eat at night before a school day, and it prefers different food on different days. Putting the correct food into its mouth makes it happy and healthy. The behavior of organizing the schoolbag and the fantasy story share similarities in the acts of putting something into a container (i.e., kinesthetic) and the spatial relation between the object and the container (i.e., books "in" the bag vs. food "in" the monster pet's mouth). The schoolbag thus corresponds to the monster pet. The correct books of different subjects (e.g., math) according to the school schedule can be mapped with the preferred food on different days. To blend the two sides, we need a knowledge frame that can enable imaginative scenarios sensibly binding together the schoolbag and the monster pet. Fortunately, fantasy novels and movies like the *Harry Potter* series provide the knowledge frame [familiars], which refer to a kind of magical spirit commonly in the form of an animal accompanying a witch. As the monster pet is a magical companion of the kid, they go to school together. In each class, the kid takes out the correct book from the monster pet's mouth. The pet feels happy with the kid. The causality between packing items and feedback from the virtual pet looks sensible. Hence, the average score in perceived causality of this idea was high (3.62).

In the fantasy story, the kid feeds the monster pet every night, and the pet feels happy. The two gradually develop rapport and friendship. The pet is grateful that the kid maintains the routine without forgetting things, while developing organization skills. This exactly matches the primary reasons for practicing the routine, which are the parents' and teachers' vision too. Table 8.15 summarizes the blends.

Summing Up

I have been interested in the connection between people's daily, mundane activities and fictional, imaginative stories for individuals' interpretation and reflection, hopefully leading to meaningful action. Finding and blending an analogous story with a target behavior into a third story should suggest new interpretations for the behavior and new reasons for taking action. The process involves two steps, namely, divergence and convergence. Imagining or recalling analogous stories results in a list of candidates, which is divergence. It is followed by comparing the candidate stories with each other and identifying the most promising one, which is convergence. For divergence, I conducted a series of design workshops over several years to explore the possibilities of identifying comparable and promising stories for different lifestyle behaviors. Drawing on theories from embodied cognition, I drafted the initial design guidelines for finding analogous stories that emphasize experientially informed similarities. Following the guidelines, workshop participants and I came up with a collection of over 100 ideas. For convergence, I selected 35 ideas

FIGURE 8.11 Screenshots of the three panels showing the routine of organizing the schoolbag (left), the analogous story of feeding the monster pet with specific food (right), and the blended third story of putting food and items into the monster pet's mouth before a school day (middle).

TABLE 8.15 The blends between the routine of organizing the schoolbag and the analogous story of feeding the monster pet, plus the convergence between the routine person's vision and the developments of the pet story, as listed in the last row

Organizing the schoolbag	Selective projection and pattern completion	Feeding the monster pet appropriately (Feed Your Bag)
	[familiars]	
PUT INTO (kid, math book, bag)	→ PUT INTO (kid, math book, broccoli, pet's mouth)	← PUT INTO (kid, broccoli, pet's
GO TO SCHOOL (kid, bag)	→ GO TO SCHOOL (kid, pet)	mouth)
CLASS (kid, math)	→ CLASS (kid, math)	
TAKE OUT (kid, math book, bag)	→ TAKE OUT (kid, math book, pet) LIKE (pet, math)	← LIKE (pet, broccoli)
	[familiars]	
PUT INTO (kid, art book, bag)	→ PUT INTO (kid, art book, sandwiches, pet's mouth)	← PUT INTO (kid, sandwiches, pet's
GO TO SCHOOL (kid, bag)	→ GO TO SCHOOL (kid, pet)	mouth)
CLASS (kid, art)	→ CLASS (kid, art)	
TAKE OUT (kid, art book, bag)	→ TAKE OUT (kid, art book, pet) LIKE (pet, art)	← LIKE (pet, sandwiches)
…	…	…
Developing the kid's organization skills	Being a magical companion, the pet feels happy for the kid's progress in organization skills development	Developing friendship with the magical companion

from the workshop collection that feature salient similarities between the analogous stories and the target behaviors. Experts from the creative sectors were then invited to independently rate each idea regarding the perceived causality between the two sides, as well as the usefulness of a proposed mixed-reality system based on the idea. The evaluation study results show that perceived causality and usefulness highly correlate in the highly rated and lowly rated groups. In-depth analyses of the structural alignments and blends of the representative ideas from the different groups further provide insights on formulating guidelines for identifying appropriate analogous stories for different routines.

In the divergence process, we look for stories analogous to the target routine. An analogous story and the routine should share similarities in varied aspects of experience. Based on observations from the workshop collection, the following components are recommended for similarity search:

- the routine acts, including postures (e.g., sitting position) and bodily actions (e.g., holding and moving something, walking up).

- the settings, including visual appearances of objects (e.g., a schoolbag vs. a pet, rice vs. snow), tactile feelings of interacting with objects (e.g., warmth on the seat), spatial relations like container (e.g., books "in" the bag vs. food "in" the pet's mouth, clothes "in" the washing machine vs. an egg "in" the incubator), verticality (rice "on" the dish vs. snow "on" the animal).
- repetitive patterns in actions or happenings, like repeatedly acquiring (e.g., skills vs. talents), exchanges, sitting and standing, washing clothes, and counting days passed.
- causal relations between actions and happenings, like repeatedly building a new house because of the old one being frail and easy to destroy, comparable to repeatedly taking medications due to the weak immune system.

In the convergence process, for each pair of a routine and an analogous story, we look for knowledge frames that enable possible and sensible scenarios where both the routine acts and the story events co-occur. The knowledge frames are usually external to both the routine domain and the story world. They can be:

- common concepts in daily life, like [neighborhood], [team project], [named trains], [public space], [accommodation], and [calendar], with which most people are familiar.
- particular concepts unfamiliar to some people, who might vaguely know the terms, such as [city trails], [observation tower], and [visual journal].
- specific knowledge in particular domains, like [running buddies], [urban farming], [chain reaction], and [blockchain technology], about which people outside the domains might have only limited understanding.
- relatively niche knowledge that some people might not know but could easily learn, such as [office cars].
- purely fictional, fantasy knowledge, yet most people know through different media, such as [familiars].

The case studies in Part II and the ideas from the workshop collection provide references for a range of knowledge frames that can be invoked for blending a routine with a story. When searching for applicable knowledge frames, usually from our memory, we can go through them, from the most common concepts, through more specific domain knowledge, to some niche or fantasy thoughts that we might have come across in the past. That means, if a common knowledge frame cannot be easily related, one might need to search and learn new or niche knowledge from different sources like the Internet. An appropriate knowledge frame should enable us to put together real-world components from the routine and fictional components from the story in an imaginative and sensible mixed-reality scenario.

Those pairs with appropriate knowledge frames invoked for the blends stay in the shortlist and we move on to further examine them regarding whether the blended third story unfolds along a direction in line with the person's vision in the routine. Regarding people's visions in different routines, we can revisit their primary reasons for practicing the routine. Among the remaining ideas in the short-list, the primary reasons for respective routines seem to span a few major general areas, including physical health or fitness, mental wellbeing, personal development, professional accomplishment, and economic or environmental sustainability. Meanwhile, different routines, though sharing primary reasons in a common area, can still be motivated by nuanced specific reasons. The following delineates some specific reasons under the major common areas:

- physical health or fitness: preparing for a marathon (running), building muscles or even entering contests (workouts), strengthening the immune system (taking medications or supplements), and avoiding prolonged sitting (focusing on desk work with breaks).
- mental wellbeing: mindfulness (eating steadily), mind stimulation (focusing on desk work with breaks).
- personal development: learning a foreign language specifically for study abroad (attending classes), developing organization skills (organizing the schoolbag).
- economic sustainability: wealth growth (DCA).

Knowing and inferring the primary reasons for practicing a routine, from the general to the specific, facilitate the development of an analogous story and the blended third story in line with one's vision. For instance, blending the workout routine with the chain tale "The Turnip" yields a third story of seeing and even helping the peasant family to pull the gigantic turnip out of the soil every day after workouts. The primary reasons for doing workouts typically include building muscles. With stronger muscles, one can help pull out the turnip more. With more muscles on the body, one might even want to enter bodybuilding contests. To further align with this specific vision, the turnip story can be elaborated into sequels, as suggested in Chapter 6, wherein the peasant grandfather also wanted to submit his gigantic root vegetables to farmers' contests. Another example of vision alignment is the blend between the stair climbing routine and hiking to enjoy spectacular views. Knowing that the primary reasons for stair climbing include not only physical health but also environmental sustainability, the blended third story can be elaborated into hiking in different areas and seeing more diverse natural environments, including *36 Views of Mount Fuji*. This story development matches one's vision regarding both physical health and environmental sustainability. Other examples include the idea of Sitravel. It blends the routine of focusing on desk work and taking breaks with the story of traveling on a train to visit attractions. The main reason for taking regular breaks is obviously to avoid prolonged sitting and a sedentary lifestyle. Yet, taking breaks and having a look at different scenic spots also provide mind stimulation, which benefits mental wellness. This alternative reason further justifies the story of Sitravel extending to cover more different attractions, like on the Tōkaidō, for refreshing the mind.

In the shortlist, all the ideas with appropriate knowledge frames invoked for the blends also pass the examination of alignments with people's visions in the routines. This shows the powerful potential of an appropriate knowledge frame invoked in an idea. As both the routine and the analogous story are intriguingly reorganized and accommodated in the frame, it provides direction and flexibility for elaborating the blended story in line with the primary reasons behind the routine. With the vision alignment, a person can practice the routine to express themselves by developing their very own third story.

9

POST-REALITY

Expressive iteration is characterized as the phenomenon of actions or behaviors that are regularly performed with subtle variations, resulting in series of tangible outcomes that elicit images and constitute narratives. The successive outcomes are often self-satisfying, because they align with a person's vision. The tangible narratives are also appealing to others, for they demonstrate possibilities of and alternatives to mundane, repetitive behaviors. Exemplars of iterative expression can be found in folklore such as chain tales across cultures (most folktales have been distributed with varied illustrations) or traditional Chinese winter calendar charts, in creative works like Monet's *Haystack* series and many notable ukiyo-e print series, in contemporary conceptual art like Sol LeWitt's serial art project. Applying expressive iteration to daily routines can make the repetitive acts self-satisfying and appealing too. The processes include first the divergence step, finding or imagining stories analogous to the target routine in the actions, settings, repetitive patterns of them or causal relations between them, followed by the convergence step, recalling or learning an external knowledge frame that informs imaginative scenarios wherein routine acts and story events can sensibly co-occur. Such scenarios typically need to accommodate both real-world components from the routine and fictional components from the story. For example, in blending the stair-climbing routine with hiking to enjoy spectacular views, the knowledge frame [observation tower] enables imagining of someone walking up the stairs in a commercial or residential building and seeing the view of a mountain probably through a window. The scenario is imaginative, because there is no actual mountain near the urban area where the building is situated. Literally, it is a blended or mixed reality, which can be simulated by today's advanced visualization technologies, commonly called "virtual reality" (VR), "augmented reality" (AR), or "extended reality" (XR). In this particular imaginative scenario, for instance, the window in the staircase can

DOI: 10.4324/9781003391449-12

be a digital display simulating a window with a virtual mountain view. For a more realistic simulation of a sense of depth in the virtual window view, or what is called "the parallax effect," some intriguing technologies are available, including cameras and sensors detecting the viewer's (i.e., the stair-climbing person) position and the display showing in real time the corresponding angle of view supposedly seen through the window. One could also consider wearing a VR/AR headset, such as Apple Vision Pro launched in 2024. It instantly captures one's physical environment (i.e., the staircase) through high-quality cameras, adds digital content (i.e., the window view), and shows the composite, stereoscopic images directly in front of the user's eyes. Some might consider wearing a headset while walking up the stairs is impractical and unnecessary, but many YouTubers have tried wearing one in the street, in a café, or on a train for hours! Wearing a similar device in daily life would probably be more common in the near future. Meanwhile, similar AR effects can be easily achieved using a mobile phone running an AR app, if one prefers. All in all, blended scenarios between routines and analogous stories are not just imaginary but can be simulated by accessible technologies of today.

When a person practices a routine (e.g., climbing the stairs), fictional components from an analogous story (e.g., a mountain view) virtually appear (e.g., through AR) in the real-world context (e.g., the staircase). The person can imagine a third scenario (e.g., walking up an observation tower in the countryside) enabled by an invoked knowledge frame and perceive the causality between the routine act and the mixed-reality outcomes (e.g., seeing a mountain through a virtual window). Since the mixed-reality outcomes align with the person's vision, and imagination affects one's intention, as discussed in Chapter 3, the person has stronger reasons to repeat the routine. This forms a feedback loop. It involves always-on, embedded, and connected sensing technologies that capture data constantly generated from human behavior, artificial intelligence (AI) techniques like machine learning that classify, aggregate, and convert raw data into information, advanced visualization technologies that simulate fictional worlds, represent the intangible information by virtual entities, and integrate them into real-world daily settings, which prompt mental simulation and likely motivate further human behavior. In short, the loop starts with human behavior, through information and computer simulation, blended into daily context, which prompts mental simulation, and finally goes back to influence human behavior (see Figure 9.1). This loop spans not only the cognitive aspects like imagination and behavior but also the socio-technological systems including real-world settings and virtual technologies. I call this "post-reality feedback loop," which has been introduced and discussed at length elsewhere.[1] In this final chapter, I show how to use the processes of the feedback loop as a guidance for designing and developing mixed-reality experiences for expressive iteration.

The post-reality feedback loop describes the reciprocal relationship of human behavior and emergent technologies including the spectrum of VR/AR/MR, digital representation of everything from the quantified self, avatars, and digital twins to the so-called "Metaverse," and generative AI. Many people see this as a sign of

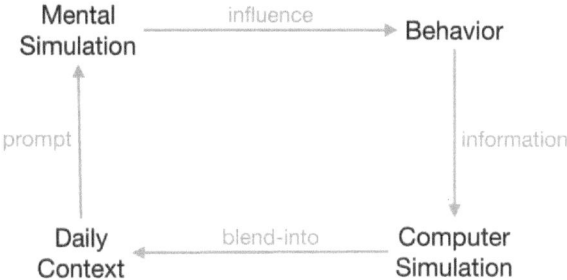

FIGURE 9.1   The post-reality feedback loop.

the rise of machines and the fall of humanity. Can we, humans, consciously take control of our behaviors and imagination in the feedback loop? Can individuals be empowered to harness emergent technologies to express iteratively in line with one's vision? With recent advances in generative AI technologies, for instance, instead of passively waiting for dystopian future scenarios entailing mass job loss, under-representation of some minorities, floods of misinformation, manipulation of choices, lack of control in privacy, and even intelligence explosion, or conversely anticipating a constantly improving, productive, and convenient lifestyle, individuals might take immediate action to make use of AI to create simulations of imaginative, self-satisfying, and appealing stories and images integrated in daily activities. Every individual could iteratively express themselves through daily practice, which represents one's unique identity. The last section of this final chapter attempts to imagine this possible third way.

The Post-Reality Feedback Loop

The term "reality" is no longer singular. While physicists might focus on objective reality, which generally refers to the mere existence of things whose properties are independent of any people or other beings who perceive and experience them,[2] some psychologists or philosophers believe that so-called "reality" is constructed with objects' properties including color, shape, odor, weight, and the like, and fabricated through our senses.[3] In the latter sense, when a person's sensory experiences of objects in an environment can be manipulated or simulated, it is likely that an alternate reality can be constructed in one's perception. Recent advances in computing and rendering technologies enable real-time simulations of varied physical properties including optics such as light diffusion, reflection, object occlusion, and stereoscopic vision, as well as dynamics like object motions, collisions, and complex movements of a crowd or hair. These simulations enable perceptions of a seemingly or "virtually" real artificial environment. VR, to some psychologists, is defined as the "externally mediated presentation of sensory stimuli" that enables perception of an "artificial environment" as non-synthetic to a certain extent.[4] The sensory stimuli may span across modalities from the visual and auditory senses, which are more easily simulated, to the tactile and olfactory senses, which have been addressed by increasingly promising technologies.[5] Although the sensory stimuli can be mediated or synthesized by different means, for example, by pre-digital media technologies like the gramophone for sounds, the eighteenth-century panorama and cyclorama for immersion,[6] or even using fans and water spray to create "special effects" in an adventure-themed roller coaster, VR today generally refers to using digital technologies to create simulations of an immersive, inter-active, and even transformative environment. The pioneering VR researcher and visionary author Jaron Lanier, based on his extensive experience in the field, pro-vocatively offers 52 different definitions of the term in his book *The Dawn of the New Everything* (2017). Among the many definitions, seven of them have the word "simulated" or "simulator." Before VR becomes commonplace today, the digital theorist N. Katherine Hayles, on the other hand, has discussed in *How We Became Posthuman* (1999) the techno-cultural interplay between virtuality, "a world of information," and materiality, the real world. To strategically define virtuality, she also starts with a common interpretation of it as "computer simulations that put the body into a feedback loop with a computer-generated image" (14). She uses the classical video game *Pong* as an example in that the game environment, including the two-dimensional "table," the paddles, and the ball, is a world of information, partly collected from a player's real-life actions and partly computed through a simulation model of the physical game mechanics. This virtual world exists in par-allel to the real world, and they intersect at many points, including the moment the player rotating the physical control knob to hit the virtual ball. This is like envision-ing the so-called "digital twins" or "omniverse" advocated by technologists and most prominently the founding CEO of the technology enterprise Nvidia.[7] While a simulated reality can precisely mirror the physical world as in the case of digital twins, it can echo part of the real world and differ in other parts for more imagina-tive and thought-provoking purposes, as in expressive iteration. Expressive itera-tion, involving cycles, also entails a kind of feedback loop similar to that between the player and the virtual environment in a game enabled by computer simulation. Yet, the feedback loop in expressive iteration extends to cover simulations not only in computers but also in the human mind.

Simulation, generally speaking, is imitation of real or hypothetical situations, including processes and possible outcomes. The imitation may be performed on a computer, which takes input data (e.g., dimensions of a paddle and a ball) and follows a model (e.g., collision detection and rebound) to compute the outcomes (e.g., the angle and speed of the ball after a hit by the paddle), followed by a visualization (e.g., a two-dimensional animation of the ball moving across the game board). Computer simulation is pervasive today, commonly found in special effects of movies, computer games, virtual worlds that enable AI robots to learn to walk on a terrain, and machine learning models that forecast the weather, among other applications. However, another kind of simulation, developed well before the advent of computers, has been taking place all along in the human mind. Mental simulation, as mentioned in Chapter 3, is defined as the imitative representation of real or hypothetical events, such as the recalling of a dinner gathering a week ago, rehearsing greeting an important guest before meeting them, or imagining the possible outcomes of choosing different forms of transportations on a busy morning.

The feedback loop connecting computer simulation and mental simulation prevails on multiple levels, from the cognitive to the socio-technological. On the cognitive level, a user interacting with a technology perceives and interprets output from the machine, and takes responsive action as the next input to the machine. The user's sensorimotor faculty is engaged in a feedback loop, which is the fundamental of what Norbert Wiener called "cybernetics" (1961) and today's VR. On the socio-technological level, people engage with various technologies in daily contexts, constituting the social and economic phenomena that further motivate technological design and development, which then reinforces technology adoption and the associated cultural perception. As Hayles (1999) puts it, virtuality is the socio-cultural perception and technological development of everyday material objects being "interpenetrated" by informational patterns (14). Meanwhile, I argue that the feedback loop also spans multiple domains, like the interconnection of computer simulation and mental simulation. In the course of computer simulation, information is generated from human behavior and visualized as computer-generated images of another world, followed by integration into a daily context. In the course of mental simulation, the unusual blend prompts our imagination, which may then lead to a change of our intention, and therefore our behavior in real life, generating new information to feed back into the computer simulation this way. Computer simulation and mental simulation therefore bring about changes, respectively, in technology-enabled simulated realities and the original reality. The changes from both sides affect each other. The latest artificial realities provide a new force to change and update human behavior and technological developments in the real world, which continues to support simulations of possible, novel realities. I call this cycle the post-reality feedback loop. In the following sections, I briefly describe the loop segment by segment.

From Behavior to Information

With integrated, connected, and always-on sensing technologies, data related to physical motions and changes inside an environment is generated in real time. The collected raw data can be aggregated as data points, which are categorized into clusters or patterns using different models and techniques for different applications. A classification model draws boundaries among data points based on values of certain features and then separates the data points into different categories.[8] The boundary condition can be static. When reading a data stream of electric resistance from a flex sensor, which measures the amount of bending on a surface, a constant threshold can be used to determine if the surface is bent over 60 degrees, and the resistance that goes up and down can be counted as a repetition of a bodily motion, say lifting a dumbbell with a bicep curl, if the sensor is stuck on one's arm. On the other hand, the boundary condition can be dynamically determined by advanced AI techniques like machine learning. With tremendous amounts of data, an AI model can be trained to find the optimal boundary condition. Consider the above flex sensor counting someone's bicep curls. The bending threshold for a count of curl can be automatically and dynamically adapted to one's own data history. When the average angle of bending gradually drops, that might be a sign of incorrect posture and one might receive an advice from the system. In sum, behavior generates data, which are constantly collected and organized into information through different classification models and mapping techniques.

The Embodiment of Information

Information is intangible structured content, which is commonly embodied through a medium or material. For example, a workout session is a piece of information consisting of the location (e.g., the gym), the time, and the repetitions of each exercise, all of which have to be represented in some form of media, whether numbers, texts, pictures, sounds, or virtual spaces. All media forms are physically embodied or "stored" in a medium made of a certain material. Numbers and texts can be inscribed on marble stones or bamboo slips, inked on papyrus, printed on paper, or displayed on an LED screen. Pictures can be painted on a rock, printed on textiles, or optically projected from film onto reflective surfaces. Sounds can be recorded and played back on vinyl records, magnetic tapes, or optical discs. Virtual spaces are rendered by both software and hardware. These media are built on the physical sciences, the advancements of which have progressed from the visible (stone, ink, paper, film) to the invisible (electromagnetic field, semiconductor). Advanced visualization technologies of today support presentation of information in the form of artificial sensory stimuli almost anywhere and at any time in daily life. Generative AI that has been rapidly developing further facilitates the creation of imaginative digital contents from flat images, three-dimensional models, animations, and videos, to even volumetric captures, which can be used to visualize any fictional worlds. In short, information is inherently intangible and yet it can be visualized and embodied in another world.

Integration into Daily Context

When a piece of information is embodied (such as in a hand gesture or an emoticon displayed in a notification), its interpretation is dependent on the context. The same hand gesture may mean "goodbye" when leaving a restaurant or "hello" when meeting someone in the Metaverse. A side view of the game character Mario running, when displayed as an emoticon in a text message from a friend on a mobile phone, might mean that the friend is coming over. The same image shown on a smartwatch when one is running, as the case mentioned in previous chapters, indicates that Mario is running along with the person as a buddy. Novel combinations of content and context stimulate imaginative interpretations.

Stimulating Imagination and Behavior

As mentioned in Chapter 2, environments have effects on individuals' behaviors because of the perception of environmental cues. This follows the thesis of dual cognitive processing, which generally includes automatic thinking and conscious thinking. At first, our perception, judgments, and actions within a new environment are all conscious processes. Yet, after some repetitions, we become habituated to the "new" environment, and those cognitive processes become increasingly automatic as the brain works to save mental effort. Changes in an environment, particularly unexpected or unusual ones, stimulate conscious thinking again. In case of expressive iteration, when digital content is integrated into daily context in a nuanced way, the unusual combination prompts association of external knowledge frames that offer imaginative yet possible scenarios in another reality. As discussed in Chapter 3, imagining a scenario frequently renders it seemingly more likely to occur. This perception of likelihood is shown to influence behavioral intention and even actions.

To sum up, design and technology enable the links between behavior and information, which is then connected to embodiment in an imaginative fictional world, via an unusual mix of that world and reality, which in turn affects the real-world behavior finally, forming a post-reality feedback loop.

Delineating Mixed-Reality Experiences

The processes of the post-reality feedback loop can be a useful framework for individuals to prepare and develop one's own expressive iteration of a routine. According to the consolidated list of routines in Table 8.4 and the remaining analogous stories after the convergence process discussed in Chapter 8, Table 9.1 delineates the design and technical considerations along the feedback loop. Some ideas in the list might require more sophisticated knowledge and skills in digital media production and computing for an ideal version. Yet, most ideas have watered-down approaches for general people to be able to materialize their expressive routines.

Human Behavior and Information

The first column of Table 9.1 is a list of the suggested technologies for collecting the relevant data and information in each routine. When a person performs a routine, wearable or mobile devices, or some Internet-of-Things (IoT) products, can be used to collect data from the routine act. In a running routine, for instance, a mobile phone, a smartwatch, or a smart bracelet can record and report data like distance traveled, average speed, and heart rate. To reduce daily tobacco consumption, a person might put an ashtray on a smart drip coffee scale that comes with a timer before they smoke. As an IoT product, the smart scale connects with a mobile app, which includes an auto-start function and automatically logs the date, time, and duration of every smoking session. The same smart scale can also be used to track the routine of drinking water regularly by putting a glass on it before one pouring oneself a glass of water.

Some routines and analogous stories require more sophisticated techniques to track bodily actions. In Stretching Wings for the stretching routine, the repetitions of the turning action determine how many times the virtual eagle turns while gliding. To track the repetitions of one's turns, one approach is to use a VR/AR headset with bodily action tracking capabilities or a mobile phone with built-in cameras and develop an automatic rep counter application,[9] which can also transform the counts instantly into digital content—the virtual eagle's movements. Another less technically challenging method can be using currently available wearable or mobile apps with an automatic rep counting feature (e.g., Gymatic Workout Tracker) and then manually converting the data into digital content afterward. In a similar fashion, other routines on the list can also generate visualized fictional outcomes after one complete cycle of the routine is logged. Hence, simply using a mobile device to log those cycles would suffice.

Computer Simulations of Fictional Worlds

After one complete routine is logged, and the data is processed into relevant information, the next step is representing the information in the corresponding analogous story. The process involves rendering the fictional components from the story as virtual items. The presentations of the virtual items should simulate how a person might perceive in the story setting. For instance, in the Super Mario Bros. story for the running routine, Mario and the person are running side by side, and so the person should see a side view of Mario. If the virtual Mario is supposed to appear on the screen of a smartwatch, a two-dimensional graphic of Mario's side profile is appropriate. If Mario is seen through a VR/AR headset, a three-dimensional computer graphic is needed. That said, the person mainly sees one side of this three-dimensional Mario. In the consolidated routine list, other analogous stories that might require three-dimensional rendering include Water Plant for drinking water and Feed Your Bag for organizing the schoolbag. In Water Plant, a person might wear a VR/AR headset and see a virtual plant "anchored" on a physical table. In Feed Your Bag, a kid might also see through a VR/AR headset a pair of monster's eyes and a pair of short hands attached to the physical schoolbag. Similar anchoring AR effects can now be achieved not only through building native mobile or VR/AR headset apps (that are usually platform-dependent) but also using web technologies (that are commonly supported across platforms and require no app installation).[10] These technologies allow anchoring virtual items to a user-selected spot on a real-world plane surface, on a flat or curved real-world image, around a recognizable physical object, or most commonly on any human face. Some of the web-based platforms even provide visual-based coding-free authoring tools. One could build a project to anchor a virtual plant on a recognizable counter in the kitchenette or overlay a pair of monster's eyes onto a printed sticker on the school-bag. To create three-dimensional virtual items, those not familiar with three-dimensional computer graphics (3DCG) tools (e.g., Blender) may use generative AI tools instead. Some of them may even generate three-dimensional models with textures from written descriptions.[11]

In other analogous stories, people practicing the routines are supposed to perceive the fictional components with more specific views, such as the window views in "The Turnip," Staircase Hiking, and Sitravel; over-the-shoulder camera views in "Tsarevitch Ivan, the Firebird and the Gray Wolf" and Stretching Wings; or typical eye-level views in "The Three Little Pigs" and "The Stonecutter." These views are relatively fixed, and so they can be simulated by two-dimensional images, and at best, enhanced with dynamic effects like parallax or flying through. Today's generative AI is able to generate from text to two-dimensional images based on a prompt that specifies the visual content and a specific angle of view and then extends it into a sequence to simulate the parallax effect and even a fly through.[12] Creating digital content to simulate the fictional components in these analogous stories becomes doable to people who are less equipped with digital media skills, because of the advancements in generative AI today.

Some other analogous stories present fictional components that are purely two-dimensional, like the screenshots of social media app user interfaces in "The Flying Ship" and "How Eleven Made Their Way in the League," or charts and patterns in Counting the Nines and Incomplete Asanoha. These two-dimensional graphics can be easily created using common graphics software tools like Illustrator, interface prototyping platforms like Figma, or even slide presentation programs like PowerPoint. In summary, the second column of Table 9.1 describes the corresponding kind of view and visual content in each analogous story.

Digital Content Blended into Daily Context

After preparing the digital content based on an analogous story, one needs to consider how to present it in the routine setting. Are any screens, mirrors, pictures, or projections, commonly found or likely to appear in the setting? Can one of them be pre-installed? In Stretching Wings and Mood Bubbles, the stretching and breathing exercises are typically performed in front of a mirror. The mirror is a medium allowing digital displays to be embedded behind it to show digital content for one to see in the front while still reflecting an image. In Smoked House, the act of smoking in an enclosed space like an office or a home might take place with a wall-hung picture nearby, which can be simulated by a digital display. Sometimes, the invoked knowledge frame for the blends between a routine and an analogous story might suggest some kind of "portal" that allows the digital content to appear in the routine setting. For instance, the knowledge frame [urban farming] in blending "The Turnip" with the workout routine, [observation tower] in Staircase Hiking, [office cars] in Sitravel, and [accommodation] in blending *53 Stages of the Tōkaidō* with the laundry routine, all include buildings, houses, or vehicles wherein windows may exist. A window in the gym, the staircase, the office, or the guesthouse can be simulated by a digital display mounted on the wall showing a virtual view outside.

In routines and corresponding analogous stories taking place in outdoor settings, such as "Tsarevitch Ivan, the Firebird and the Gray Wolf" for the commute routine, *100 Famous Views of Edo* for the takeout lunch routine, it is not easy to pre-install digital displays or projections. Wearing a VR/AR headset to see digital content, such as Ivan riding the Gray Wolf or the view of Edo, blended in physical routine settings, like in a train compartment or on a rooftop, is ideal. Meanwhile, some people might not see wearing a headset appropriate for outdoor use. Using a mobile device with built-in cameras and a mobile or web AR app can achieve similar results. The third column of Table 9.1 lists out the device or installation required to blend digital content in the routine setting for each case.

TABLE 9.1 A list of routines with promising analogous stories whose mixed-reality experiences are delineated according to the processes in the post-reality feedback loop

Human behavior and information	Computer simulation of fictional world	Digital content blended into daily context	Mental simulation based in knowledge frames
Goal-directed routine			
Running Mobile or wearable devices: step counts, distance traveled, average speed, and heart rate	Super Mario Bros.: A side view of Mario, or a three-dimensional rendering of him, who is running	A smart watch app displaying a side view of Mario running; or a VR/AR headset showing a three-dimensional Mario running next to me.	[Running buddies] Mario and I are running buddies. We run side by side every morning in a training program.
Workouts Mobile phones: logs	The Turnip: A window view on a farm, including a gigantic turnip, grandpa, grandma, granddaughter, ...	In the gym, a wall-mounted digital display showing a virtual window view; or a mobile AR app showing a virtual turnip in the park near the gym.	[Urban farming] In a farm near the gym, I, as a strong person, help the peasant family pull the turnip out of the soil.

(Continued)

TABLE 9.1 (Continued)

Human behavior and information	Computer simulation of fictional world	Digital content blended into daily context	Mental simulation based in knowledge frames
Workouts Mobile phones: logs	The Fallen Tree: A window view on a trail blocked by a large fallen tree, and some volunteers working to clear the path.	In the gym, a wall-mounted digital display showing a virtual window view; or a mobile AR app showing a virtual fallen tree in the park near the gym.	[City trails] In a park near the gym, I, as a strong person, volunteer to help clear the path.
Taking medications Mobile phones: logs	The Three Little Pigs: An eye-level view of a house built by the little pigs being blown down by the wolf	A tablet app showing a virtual camera view of a neighboring house being blown by the wolf	[Neighborhood] I feel unwell because it is windy outside. Taking medicine helps me build strong immunity. I feel better instantly and help my neighbors rebuild their houses.
Attending classes Mobile phones: logs	The Flying Ship: An eye-level view of a talented person showing his special skill	After each class, a mobile app shows a new virtual member joining the project's social media group and posting videos about his skill.	[Team project] I have come across and invited a talented classmate to join the project. I learn something from them.
Attending classes Mobile phones: logs	How Eleven Made Their Way in the League: An eye-level view of a talented soccer player showing his special skill	After each class, a mobile app shows a new virtual member joining the soccer team's social media group and posting videos about his skill.	[Team project] I have come across and recruited a talented classmate to join the soccer team. I learn something from them.
Commuting to do piecework Mobile phones: location-based sensing like assisted GPS	Tsarevitch Ivan, the Firebird and the Gray Wolf: An over-the-shoulder view of Ivan, who is riding the Gray Wolf	On a moving train, a mobile navigation app displaying a virtual character Ivan riding a virtual wolf moving on a map in real time	[Mission titles] [Named trains] I have to pursue the Firebird. I ride a train named "Gray Wolf." I see myself riding it and moving on the navigation map.

(Continued)

TABLE 9.1 (Continued)

Human behavior and information	Computer simulation of fictional world	Digital content blended into daily context	Mental simulation based in knowledge frames
Trading used items Mobile phones: logs	The Stonecutter: An eye-level view of the stonecutter transforming into a rich man	After a trade at a store, a mobile app showing a virtual stonecutter's transformation.	[Chain reaction] I trade a used item for a new good, and the action triggers chain reactions, leading to the stonecutter making a wish and being transformed into a rich man.
Trading used items Mobile phones: logs	The Chef: An eye-level view of the chef leaving a restaurant in Japan and moving to a new one in France	After trading a used item at a store, a mobile app showing a virtual chef's relocation for a new job.	[Chain reaction] I trade of a used item for a new good, and the action triggers chain reactions, leading to the chef's decision of joining a new restaurant.
Dollar cost averaging Mobile phones: logs	*Incomplete Open Cubes*: A new variation of incomplete open cubes	After an online stock purchase transaction, a mobile app simulates a composition of three rods into a newly found incomplete open cube and logs it.	[Blockchain technology] I have made a stock purchase. The transaction is confirmed and the record is appended to the distributed ledger with an incomplete open cube as proof-of-work.
Dollar cost averaging Mobile phones: logs	Incomplete Asanoha: A new variation of incomplete asanoha	After an online stock purchase transaction, a newly found incomplete asanoha is added to the projection of a hemp field in the sitting room.	[Blockchain technology] I have made a stock purchase. The transaction is confirmed and the record is appended to the distributed ledger with an incomplete asanoha as proof-of-work.

(Continued)

TABLE 9.1 (Continued)

Human behavior and information	Computer simulation of fictional world	Digital content blended into daily context	Mental simulation based in knowledge frames
Reducing tobacco consumption IoT products: smart scale	Smoked House: Dust building up on a flat surface over time.	A wall-mounted digital picture frame shows a family photo, which is gradually covered by virtual dust.	[Air pollution] I consume cigarettes consecutively. The smoke drifts near the picture frame. Dust gradually builds up and covers the family photo.
Maintenance routine			
Drinking adequate water IoT products: smart scale	Water Plant: A three-dimensional rendering of a plant getting refreshed and growing over time	Through a VR/ AR headset, a three-dimensional virtual plant is added on a physical table. The virtual plant gradually grows.	[Humidity] Due to low humidity, both the plant and I need water. I pour water into a glass and on the plant.
Taking stairs Mobile or wearable devices: flights climbed	Staircase Hiking: A window view of a mountain seen from somewhere at a certain altitude	In the staircase of a building, a wall-mounted digital display showing a virtual window view	[Observation tower] I walk up the stairs to my office in a building. Through the window on every floor, I see the mountain outside. As I reach the floor where my office is, I see the mountain top.
Taking stairs Mobile or wearable devices: flights climbed	*36 Views of Mt. Fuji*: A window view of Mount Fuji seen from somewhere at a certain altitude	In the staircase of a building, a wall-mounted digital display showing a virtual window view	[Observation tower] I walk up the stairs to my office in a building. Through the window on every floor, I see Mount Fuji outside. As I reach the floor where my office is, I see the top of Mount Fuji.

(Continued)

TABLE 9.1 (Continued)

Human behavior and information	Computer simulation of fictional world	Digital content blended into daily context	Mental simulation based in knowledge frames
Having lunch Mobile: logs	*100 Famous Views of Edo*: Different views of the Edo city at a particularly location	In a park, a mobile AR app showing the physical meal with a virtual background of the Edo city.	[Public space] I have bought a takeout lunch and walked to a park. I use the mobile phone to take a snapshot of the meal. The background of the shot tells me that it is somewhere in Edo.
Focusing on desk work with breaks Mobile or wearable devices: sit hours	Sitravel: A window view of an attraction seen from a train station	In an office, a wall-mounted digital display showing a virtual view of an attraction through a window	[Office cars] I focus on my desk work for an hour. I stand up to take a break. The window view tells me that the train I am riding has already arrived at a station near an attraction.
Doing laundry at home Mobile devices: logs	*53 Stages of the Tōkaidō*: A road scene near a station on the Tōkaidō	In a home setting consisting of a washing machine, a wall-mounted digital display showing a virtual window view of a road scene	[Accommodation] After doing my laundry, I am ready to take off. The window view shifts to another road scene, showing that I have arrived at another station on the Tōkaidō and will stay in another guesthouse.
Walking the dog Mobile devices: logs	Counting days on a nine-nines chart: One stroke of a character being written on a chart	In a home setting, a wall-mounted digital display showing a virtual chart	[Visual journal] I walk the dog in the park and then we go back home. I try to visualize the grass I have seen along the path. One stroke is drawn every day. After 81 days, I complete nine pictures and give the scenery I have seen a caption.

(*Continued*)

TABLE 9.1 (Continued)

Human behavior and information	Computer simulation of fictional world	Digital content blended into daily context	Mental simulation based in knowledge frames
Stretching VR/AR headset: motion detection; or mobile device: logs	Stretching Wings: A view taken over the shoulder of an eagle, which glides over a landscape and turns left and right repeatedly	In a home setting, a smart mirror (a LED display behind a reflective glass) showing a view taken over the shoulder of a virtual eagle.	[Giant bird riding] I turn left and right repeatedly in my stretching routine. I see in the mirror that I am riding a giant bird that glides over a landscape to find food and carry back to feed its chicks in the nest.
Breathing exercise Mobile devices: logs	Mood Bubbles: Different bubbles constantly coming out from the center and moving toward the viewer	In a home setting, a smart mirror (with a LED display behind a reflective glass) showing different virtual bubbles constantly appearing	[Mental imagery] I sit down on the floor and breathe slowly. I see bubbles coming out from my nose in the mirror, when I exhale. I focus on my breath and the bubbles I blow.
Organizing the schoolbag Mobile devices: logs	Feed Your Bag: A three-dimensional rendering of an adorable monster with eyes and short hands	Through a VR/AR headset, a pair of virtual monster eyes is added on a physical schoolbag. The virtual monster shows emotions through its eyes and hands.	[Familiars] I prepare the books and items for tomorrow's classes. I put them into my monster bag's mouth. He feels satisfied. When I have the correct books and items in a class, he is happy for me.
Self-monitoring peak flow Mobile devices: logs	Peak Flow Flute: A section of flute music being played	Through a smart speaker paired with a peak flow meter, a music piece can be played.	[Music recording] I blow into the peak flow meter, which is also a smart flute. It records the section I just practiced and plays it back. A new section is recorded every day. They form a whole music piece.

Mental Simulation Based in Knowledge Frames

Finally, in the fourth column, I take the perspective of the person who performs the routine and articulate what one might perceive and imagine.

Identity in Post-Reality

Expressive iteration is a phenomenon observed in the practices of many creative individuals. Monet repeatedly attempted to capture the diffusion of light at different instants; ukiyo-e masters repeatedly created and published woodblock print series as visual documentaries of the prosperous Edo period; Sol LeWitt enumerated all possible configurational variations of an incomplete cube in authentic and pure manner according to his self-proposed plan and rules; numerous unsung heroes among common people across cultures invented, revised, and retold varied chain tales that give color to their repetitive daily activities; commoners and royal members in the Imperial China who enduringly counted and documented the winter days in the forms of poems or blossom pictures to help themselves survive the bitter winters with hope. Instead of iterating to fix problems and pursue the final satisfying outcomes, they iterate to express themselves through the processes. They embody the characteristics of expressive iteration. They practice repetitive acts and perceive self-satisfying outcomes that show progression and variations continuingly in line with their visions. Their practices result in tangible narratives that demonstrate appealing possibilities and alternatives in mundane daily lives. They are exemplars of expressive iteration and references for us to make routines more meaningful.

Today, our daily lives and the post-reality feedback loop are inextricable. The loop spans both technology-enabled artificial realities and our original reality, which are constantly updated and shaped by machines through computer simulation and humans through mental simulation. With the continuing advancements of machine intelligence and computer graphics, computer simulation may take a pivotal role in mobilizing the loop, which will pervade our daily lives. The computer simulation system can pick up information on human behaviors and project different possible realities to engage more people and keep them interested. This is likely to be the primary objective of many service providers or technology enterprises, which are typically behind the computer simulation system. The companies running video streaming or short video platforms, computer or mobile games, car-hailing or food delivery applications, numerous AI-powered productivity applications, and recently self-driving vehicles, for instance, would like to collect more behavioral information of their users and turn them into more personalized simulated experiences that permeate their daily lives. These technologies and designs seem to ultimately keep the users in the feedback loop for these companies.

Worse still, the impacts of AI have attracted major public attention since the launches of AI products like ChatGPT that seem to outperform human intelligence in not only goal-specific tasks like data analysis, speech recognition, translation, and wayfinding, but also creative practices like story and poetry writing, music composition, image and video making, and even computer programming. In addition to the bias in the data and algorithms used to train AI models, underrepresentation of minority groups or species, moral decision making integrated in AI, malicious use of AI for creating and spreading misinformation, personal data privacy issues, and job displacement in various sectors, people are also concerned about copyrights issues of AI-generated content and potential devaluation of creative works. Now, people can use generative AI tools to create stories, images, music, speeches, videos, animations, and three-dimensional models for computer graphics, in a much more convenient way than before, making the production of creative works seemingly manageable on the scale of a small group or even an individual.[13]

Some people might think that this would lead to floods of content in the market, loss of authentic value of human creativity, and ultimately identity crisis of humanity. Facing the constantly advancing AI technologies, some visionaries started to advocate the use of new metaphors to see AI, changing from an intelligence that imitates humans,[14] to a new digital species,[15] or more theoretically a new cognitive entity in addition to humans and non-human animals, as Hayles (2023) puts it. A cognitive entity, in Hayles's sense, is able to access and interpret information, and so engages in some kinds of behaviors, which refer to not only animal behaviors, but also biological processes in plants, such as photosynthesis, osmosis, and pollination, as well as computational algorithms in AI. In this sense, a plant is a cognitive entity that can "interpret" changes in its environment, like sunrise and sunset, temperature, humidity, and soil fertility, and then perform different behaviors accordingly, such as blossoming and bearing fruit. An AI is also a cognitive entity that learns from training data, interprets new input data, and then takes action according to its algorithms. Meanwhile, each cognitive entity engages in its own relevant context. A person of course has to deal with many aspects of life. A plant only attends to the atmosphere, the soil, and probably a small group of other co-inhabitants. A wild animal might pay attention to a broader habitat and most importantly, other preys and predators. An AI focuses on all the available data and deals with over millions of parameters in its model. The difference of each cognitive entity's context from others renders one's unique characteristics. Some pine trees in California have a lifespan of over a thousand of years, but they only attend to and engage in their neighborhoods. Their tree rings thus can tell long climate history in their forest area.[16] Some birds migrate long distances crossing continents and oceans every year for food and breeding, yet they only sense the subtle climate change over a few decades.[17] A specially trained GPT (Generative Pre-trained Transformer) is able to access data and information across the globe about climate change, and it might be able to live long enough and evolve to accurately predict the weather at anywhere and anytime. We humans, typically living for several decades, are not comparable to pine trees in terms of age; we intermittently travel to different places on the Earth, but can never be as instinctive as some migrant birds; now we face challenges from AI in terms of the capacity and performance of structuring information, detecting patterns, making predictions, even creating art and framing problems. What are the remaining unique characteristics that make us humans?

Here, I see a promising way to construct one's identity with expressive iteration. One unique aspect of humans that differs from other cognitive entities including AI is our daily living. Every individual human engages in many different activities every day, including routines. Apart from the usual routines like brushing teeth, having a morning coffee, or taking a shower, some of us might have a morning running routine, walking our dogs in the evening, and reading books for an hour every night before going to bed, among many others. Living every day at a certain pace, engaging in certain daily activities, applying expressive iteration to routines, and creating tangible narratives that reflect and support one's vision is making a unique statement that other people cannot replicate, because every individual is living their life and experiencing the world differently. Even though an AI would be given a physical body very soon and become able to have multimodal senses, such an AI robot cannot live and experience the same way a typical human does. Humans have to eat, drink, and sleep to maintain basic physiological needs. Humans also grow, age, sometimes get sick, and die. Hence, we invent and practice different activities and routines. AI robots have no actual reasons for performing routines like taking a shower, doing yoga, or meditating, not to mention eating and drinking. Their sense of time passing would be very different from ours.

As mentioned in this chapter, it is not technically challenging today for a person even without technical skills in digital media to learn and use generative AI tools, web-based VR/AR platforms, and a few accessible gadgets to create mixed-reality experiences for one's own routines. To some people, integrating emergent AI and VR/AR technologies into daily routines too deeply might risk being trapped in the feedback loop. In fact, the latest developments in these emergent technologies have opened up possibilities for individuals to harness the power of the tools and create content for their own stories, instead of sticking with a product or service provided by one company for some pre-scripted user experiences in a feedback loop. Any person can be more like a hacker now, piggybacking on existing products, extracting data or information, learning to use AI tools to create and modify content, and customizing real-world settings for digital content presentation. Most importantly, we might co-imagine with AI and develop stories analogous to routines. A daily routine becomes a story reflecting one's daily life with a vision.

Applying expressive iteration, you are what you iterate.

NOTES

Chapter 1

1. See Sharp and Macklin (2019). They see iteration as a powerful way that can turn failure into creative outcomes.
2. See Groom and Shaw (2014, para 11), https://publications.artic.edu/monet/reader/paintingsanddrawings/section/135598/p-135598-11, accessed 16 July 2024.
3. https://www.theverge.com/2021/3/11/22325054/beeple-christies-nft-sale-cost-everydays-69-million, accessed 16 July 2024.
4. See Rubin (2015, 201).
5. See, for example, Wood (2019, 116–117).
6. See Bargh and Chartrand (1999).
7. See 李文君 (2006) and 海上 (2005, 163).
8. https://www.nytimes.com/2022/01/31/business/media/new-york-times-wordle.html, accessed 16 July 2024.
9. For a classification of motivation, see Ryan and Deci (2000a).
10. https://reversle.net/, accessed 16 July 2024.
11. For an introduction to self-determination theory, see Ryan and Deci (2000b). For novelty proposed as a new candidate of basic psychological needs, see González-Cutre and colleagues (2016, 2020).
12. For an introduction to gamification, see Deterding and colleagues (2011). For overviews of gamification in areas like health and education, see King and colleagues (2013) and Reiners and Wood (2015).
13. See Mekler and colleagues (2017).
14. Rapp (2017) suggests using stories as one of the design strategies for behavior change technologies. Nicholson (2015) includes stories as one of the elements of meaningful gamification.
15. See Ludden and colleagues (2015).
16. For an introduction to mental simulation, see Taylor and colleagues (1998).
17. See Green and Brock (2000).
18. See Green (2004).
19. See Nicholson (2015, 8).
20. See Rubin (2015, 199–200).

21. See Clear (2018, 192).
22. For an introduction to conceptual metaphor, see Lakoff and Johnson (2003).
23. See Sherman and colleagues (1985).
24. https://blog.fitbit.com/fitbit-badges/, accessed 31 July 2022.
25. See, for example, Baregheh and colleagues (2009) and McKinley and colleagues (2014).
26. See Griffith and Rubera (2014).
27. Banathy (1996) is arguably the first model systematically representing the usual design processes of divergence and convergence in an iterative fashion.

Chapter 2

1. See Ramage and colleagues (2016).
2. See, for example, a long-running show in Hong Kong, Good Night Show - King Maker, https://en.wikipedia.org/wiki/Good_Night_Show_-_King_Maker, accessed 16 July 2024.
3. See Bogost (2015).
4. See Fogg (2009).
5. See Oinas-Kukkonen (2013).
6. See, for example, Martin and Hanington (2012).
7. For an introduction to user-centered design, see Norman (2002).
8. See Gaver and colleagues (2004).
9. For design semantics, see Krippendorff (2006). For metaphorical meaning of design, see Rompay (2008).
10. For an introduction to usability, see https://digital.gov/topics/usability/, accessed 16 July 2024.
11. Psychologists call this dual-process theories, see Chaiken and Trope (1999).
12. For details of automatic processes, see Bargh and Chartrand (1999).
13. For an introduction to positive psychology, see Seligman and Csikszentmihalyi (2000).
14. See Ryan and Deci (2000b).
15. For the relationship among habit, planned behavior, and automaticity, see Verplanken and Aarts (1999).
16. See Chaiken and Trope (1999).
17. For an introduction to the magic circle, see Salen and Zimmerman (2004, 94–99). For the relationship between video games and real-world rules, see Juul (2005).
18. See https://sparkful.app/fortune-city, accessed 16 July 2024.
19. See Deterding and colleagues (2011).
20. See Bogost (2013) for his criticism on gamification.

Chapter 3

1. See Wardrip-Fruin and Harrigan (2004) for writings and responses between pairs of varied scholars.
2. Joe Sachs, "Aristotle: Poetics," The Internet Encyclopedia of Philosophy, ISSN 2161-0002, https://iep.utm.edu/, accessed 16 July 2024.
3. Joe Sachs, "Aristotle: Poetics," The Internet Encyclopedia of Philosophy, ISSN 2161-0002, https://iep.utm.edu/, accessed 16 July 2024.
4. Joe Sachs, "Aristotle: Poetics," The Internet Encyclopedia of Philosophy, ISSN 2161-0002, https://iep.utm.edu/, accessed 16 July 2024.
5. Christopher Shields, "Aristotle's Psychology", The Stanford Encyclopedia of Philosophy (Winter 2020 Edition), Edward N. Zalta (ed.), https://plato.stanford.edu/archives/win2020/entries/aristotle-psychology/.
6. Martin and colleagues (1999) conducts a review of over 200 scientific studies on imagery use by athletes, which finds that imagery in most cases improves sport performance.

7. Hall and colleagues (1998), through literature review, evaluation of other imagery questionnaires, and expert evaluations of research professionals and elite athletes, develops the Sport Imagery Questionnaire (SIQ), which categorizes imagery use by athletes, including cognitive and motivational functions. Beauchamp and colleagues (2002) uses a modified version of SIQ to assess both functions of imagery use by 51 professional golfers and finds that motivational imagery use mediated the relationship between self-efficacy and performance.
8. See Taylor and colleagues (1998).
9. See Pham and Taylor (1999).
10. See Escalas and Luce (2004).
11. See Pham and Talyor (1999).
12. See Chan and Cameron (2012) and Kim and colleagues (2012).
13. See Armitage and Reidy (2008).
14. See Knäuper and colleagues (2011).
15. See Conroy and colleagues (2015).
16. See Wynd (2005).
17. See Worth and colleagues (2005).
18. See Bailenson (2018, 83).
19. See Yee and Bailenson (2007).
20. See Yee and Bailenson (2006) and Oh and colleagues (2016).
21. See Hershfield and colleagues (2011).
22. See Ahn and colleagues (2016).
23. See Chow (2021b).
24. See Holland (2008).
25. See Appel and Richter (2007).
26. See Chow (2021b).

Chapter 4

1. This may be a reductive narrative. For more details about neurons in the brain from the perspective of cognitive science, see Feldman (2008). For recent advances in understanding of the brain informed by neuroscience, see https://news.mit.edu/2015/brain-strengthen-connections-between-neurons-1118, accessed 16 July 2024.
2. See, for example, Hobeika and colleagues (2016) and Whitaker and colleagues (2018).
3. See also Chatman (1978, 45–46) for difference in degree of explicitness regarding causality in narrative.
4. For more details about definition of frame, see Fillmore (1985), Lakoff (1987, 68), Coulson (2001, 19–20), and Fauconnier and Turner (2002, 104).

Chapter 5

1. For an update on Herzberg's work, see https://hbr.org/2003/01/one-more-time-how-do-you-motivate-employees, accessed 16 July 2024.
2. See Ritchie (2017, 117–118).
3. See Ritchie (2017, 113).
4. See Propp (2012, 47).
5. See Propp (2012, 48–49).
6. See Propp (2012, 39).
7. See the ATU classification presented by University of Missouri Libraries, https://libraryguides.missouri.edu/c.php?g=1039894&p=7610331, accessed 16 July 2024.
8. See Propp (2012, 40).

9. The Russian lacquer miniature, originated circa 1800, turned from the art of icon painting into depicting scenes from folktales since the early twentieth century. See, Margarita (2011, 3–4).
10. See Salter (2006, 14).
11. Based on an introduction in a publication by the Hokusai-kan Museum in Obuse, Japan, opened in 1976.
12. See https://apps.apple.com/jp/app/%E6%9D%B1%E6%B5%B7%E9%81%93%E4%BA%94%E5%8D%81%E4%B8%89%E6%AC%A1%E6%AD%A9%E8%A8%88/id1533643626, accessed 16 July 2024.
13. See https://blog.fitbit.com/fitbit-badges/, accessed 16 July 2024.

Chapter 6

1. See a related study in Raimondo and colleagues (2013).
2. See Herman and colleagues (2005, 527).
3. The term "anthology," in its Greek origin, literally means "gathering flowers" and symbolizes a collection of poetry.
4. See 潘力 (2022, 283).
5. See online archives of the prints "Shōno" in *53 Stages of the Tokaido*, https://commons.wikimedia.org/w/index.php?curid=3206382, accessed 16 July 2024, and "Suhara," in *69 Stages of the Kisokaidō*, https://commons.wikimedia.org/w/index.php?curid=2964394, accessed 16 July 2024, both of which depict similar road scenes of people caught in rainstorms.
6. See online archives of the prints "Aki Province" in *Famous Views of the Sixty-odd Provinces*, https://commons.wikimedia.org/w/index.php?curid=41356532, accessed 16 July 2024, and "Suidō Bridge at Suruga Terrace" in *100 Famous Views of Edo*, https://commons.wikimedia.org/w/index.php?curid=3790905, accessed 16 July 2024, both of which feature prominent objects in the foreground, respectively part of a torii gate and a koinobori (traditional koi-shaped windsock), in front of distant views.
7. Hokusai designed and printed an earlier series on famous places in Kyoto but it is in the traditional landscape format.
8. See Herman and colleagues (2005, 107).
9. See Zeigarnik (2000).
10. See Slowik (2014).
11. See Ryan (1990).
12. See Herman and colleagues (2005, 134).
13. See Herman and colleagues (2005, 186).
14. See Ritchie (2017, 31).
15. See Bickford (2005) and 胡櫨文 (2016).
16. See 傅連仲 (1982) and Bickford (2005).
17. It was also marked "one" at the top-right corner of the cell and "nine" at the top-left corner to mean that this cell was the first "nine." The dates of the beginning and end of this nine-day period were written on the bottom-left and bottom-right corners respectively.
18. See 海上 (2005, 163).
19. See O'Sullivan (2019).
20. See Herman and colleagues (2005, 527).
21. See Groom and Shaw (2014, para 10).
22. See https://www.tate.org.uk/art/art-terms/c/conceptual-art, accessed 16 July 2024.
23. See https://www.moma.org/collection/works/81533, accessed 16 July 2024.
24. See https://www.tate.org.uk/art/art-terms/m/minimalism and https://www.moma.org/collection/terms/minimalism/serial-forms-and-repetition, accessed 16 July 2024.
25. See https://www.metmuseum.org/art/collection/search/691091, accessed 16 July 2024.

26. Many websites revisit the work and re-present the concept in a digital fashion, including https://cubes-revisited.art/about/ and https://www.alisonpitt.com/blog/an-exploration-of -incomplete-open-cubes, accessed 16 July 2024.

27. See Rozhkovskaya and Reb (2015).

28. The term was first coined in Graham (1965). For an introduction, see, for example, https:// www.forbes.com/advisor/investing/dollar-cost-averaging/, accessed 16 July 2024.

29. For a bite-sized introduction to blockchain, see IBM website https://www.ibm.com/topics/ blockchain and 3Blue1Brown video https://www.youtube.com/watch?v=bBC-nXj3Ng4, accessed 16 July 2024.

Chapter 7

1. See Ross (1979) and Western (1980).

2. See Ross (1979).

3. See Western (1980).

4. See https://www.latimes.com/archives/la-xpm-1998-oct-25-tm-35829-story.html, accessed 24 July 2024.

5. See Stets and Burke (2000).

6. See Land (2013).

7. See https://www.netflix.com/tw-en/title/81342639, accessed 24 July 2024.

8. See https://www.fs.usda.gov/detail/giffordpinchot/workingtogether/volunteering/?cid=s telprdb5181783, accessed 24 July 2024.

9. See Salter (2006, 165).

10. One can browse the digital collections of Japan National Diet Library, https://dl.ndl. go.jp/en/pid/1311899, or search in Tokyo Metropolitan Library, https://archive.library. metro.tokyo.lg.jp/da/searchDetail?category=%E5%8F%8C%E5%85%AD, accessed 24 July 2024.

11. See Salter (2006, 165).

12. See in the description (Japanese) of a remake version, https://shop.okunokaruta. com/?pid=58152698, accessed 24 July 2024.

13. See an online archive of *Ukiyo Dōchū Hizakurige Comic Sugoroku*, https://dl.ndl.go.jp/ pid/1310717, accessed 24 July 2024.

14. See Salter (2006, 17). I also saw a similar description of the physical print once exhibited in Money and Ukiyo-e Museum, Nagoya, Japan, in January 2024.

15. See an online archive of *Famous Places of Edo Ichiran Sugoroku*, https://archive.library. metro.tokyo.lg.jp/da/detail?tilcod=0000000004-00000016, accessed 24 July 2024.

16. See Salter (2006, 16).

17. See https://www.sciencefocus.com/science/fitness-app-step-count, accessed 24 July 2024.

18. See 胡櫨文 (2006) or an online digital version, http://dlib.cafa.edu.cn/rubbing/ item/5055-n-a?offset=1, accessed 24 July 2024.

19. See Bickford (2005, 367–368).

20. See Bickford (2005, 351–353).

21. See https://www.lcsd.gov.hk/en/green/blossoms.html, accessed 24 July 2024.

22. See the original web-based version of Townscaper, https://oskarstalberg.com/ Townscaper, and the online converter, https://tarmo888.github.io/Wordle2Townscaper/, accessed 24 July 2024.

23. See Groom and Shaw (2014, para 10), https://publications.artic.edu/monet/reader/ paintingsanddrawings/section/135598/p-135598-10, accessed 24 July 2024.

24. For an introduction to asanoha and traditional Japanese patterns, see https:// www.nippon.com/en/japan-data/h00478/. For the auspicious meaning of asanoha, see

https://kumikowoodworking.com/products/category/auspicious-omens-motifs/, accessed 24 July 2024.

25. See https://www.toyodamk.com/kanuma-kumiko and https://kumikowoodworking. com/products/tn-101/, accessed 24 July 2024.

26. See https://www.toyodamk.com/product?lightbox=dataItem-kjxzgs4p3, accessed 24 July 2024.

Chapter 8

1. See Banathy (1996).
2. See, for example, https://www.medminder.com/, accessed 24 July 2024.
3. See Chow (2021b).
4. For general people's interpretations of tracked data in daily life, see Rapp and Cena (2016).
5. See Chow (2021a).
6. See Lakoff and Johnson (1999).
7. See Chapter 3.
8. For definitions of creativity, see Sawyer (2012, 8).
9. For CAT, see Amabile (1982).
10. See Sawyer (2012, 41).
11. See the introduction to design in Chapter 2.
12. See Shrout and Fleiss (1979).
13. See Mandler (1992).
14. See https://www.nhs.uk/live-well/exercise/why-sitting-too-much-is-bad-for-us/, https://www.ccohs.ca/oshanswers/ergonomics/sitting/sitting_overview.html, or https://edition.cnn.com/2024/01/25/health/effects-of-sitting-on-mortality-wellness/index.html, accessed 24 July 2024.
15. See https://www.japantimes.co.jp/news/2021/02/01/business/bullet-train-office/ and https://www.japantimes.co.jp/news/2021/11/25/business/corporate-business/jr-east-bullet-train-office-cars-telework/, accessed 24 July 2024.

Chapter 9

1. See Chow (2023).
2. See Lakoff and Johnson (2003, 186–187).
3. Donald D. Hoffman is a prominent figure who advocates this claim. See Hoffman (1998).
4. See Ryan and colleagues (2019, 18).
5. See, for example, https://cutecenter.nus.edu.sg/projects/virtual-cocktail.html and https://emerge.io/about/, accessed 24 July 2024.
6. See Grau (2003, 56–65) for the media archeology of immersive virtual art.
7. See relevant demos in https://www.youtube.com/watch?v=-eiL9HzvJh0 or https://www.youtube.com/watch?v=RGEscRUXne8, accessed 24 July 2024.
8. See an interactive scroll-based animation illustrating the idea of machine learning, http://www.r2d3.us/visual-intro-to-machine-learning-part-1/, accessed 24 July 2024.
9. See demos in https://quickpose.ai/testflight-demo/, accessed 24 July 2024.
10. See, for example, https://www.8thwall.com or https://mywebar.com/, accessed 24 July 2024.
11. See, for example, https://lumalabs.ai/genie?view=create, accessed 24 July 2024.
12. For recent research, see https://research.google/blog/infinite-nature-generating-3d-flythroughs-from-still-photos/, accessed 24 July 2024.

13. An artist used AI to create a music video and won in a competition, https://ew.com/ai-wins-pink-floyd-s-dark-side-of-the-moon-video-competition-8628712, accessed 24 July 2024.
14. In the famous Turing test for machine intelligence, named after Alan Turing, a human being should be unable to distinguish the machine from another human being by using the replies to questions put to both.
15. See Mustafa Suleyman's TED Talk, https://www.ted.com/talks/mustafa_suleyman_what_is_an_ai_anyway?subtitle=en, accessed 24 July 2024.
16. See https://climate.nasa.gov/news/2540/tree-rings-provide-snapshots-of-earths-past-climate/, accessed 24 July 2024.
17. See https://www.nature.com/articles/s41467-020-19256-0, accessed 24 July 2024.

BIBLIOGRAPHY

Afanas'ev, Aleksandr. (1916). *Russian Folk-Tales* (Leonard A. Magnus, Trans. 2nd ed.). London; New York: Kegan Paul, Trench, Trubner & Co.; E.P. Dutton & Company.

Ahn, Sun Joo (Grace), Bostick, Joshua, Ogle, Elise, Nowak, Kristine L., McGillicuddy, Kara T., & Bailenson, Jeremy N. (2016). Experiencing nature: Embodying animals in immersive virtual environments increases inclusion of nature in self and involvement with nature. *Journal of Computer-Mediated Communication, 21*(6), 399–419.

Ajzen, Icek. (1985). From intentions to actions: A theory of planned behavior. In Julius Kuhl & Jürgen Beckman (Eds.), *Action-Control: From Cognition to Behavior* (pp. 11–39). Heidelberg: Springer.

Amabile, Teresa M. (1982). Social psychology of creativity: A consensual assessment technique. *Journal of Personality and Social Psychology, 43*(5), 997–1013.

Anderson, Craig A. (1983). Imagination and expectation: The effect of imagining behavioral scripts on personal intentions. *Journal of Personality and Social Psychology, 45*(2), 293–305.

Appel, Markus, & Richter, Tobias. (2007). Persuasive Effects of Fictional Narratives Increase Over Time. *Media Psychology, 10*, 113–134.

Armitage, Christopher J., & Reidy, John G. (2008). Use of mental simulations to change theory of planned behaviour variables. *British Journal of Health Psychology, 13*, 513–524.

Arnheim, Rudolf. (1969). *Visual Thinking*. Berkeley: University of California Press.

Bailenson, Jeremy. (2018). *Experience on Demand: What Virtual Reality Is, How It Works, and What It Can Do*. New York: W. W. Norton & Company.

Bailey, Jakki O., Bailenson, Jeremy N., Flora, June, Armel, K. Carrie, Voelker, David, & Reeves, Byron. (2015). The impact of vivid messages on reducing energy consumption related to hot water use. *Environment and Behavior, 47*(5), 570–592.

Banathy, Bela H. (1996). *Designing Social Systems in a Changing World*. New York: Springer.

Baregheh, Anahita, Rowley, Jennifer, & Sambrook, Sally. (2009). Towards a multidisciplinary definition of innovation. *Management Decision, 47*(8), 1323–1339.

Bargh, John A., & Chartrand, Tanya L. (1999). The unbearable automaticity of being. *American Psychologist, 54*(7), 462–479.

Barthes, Roland. (1977). Rhetoric of the image (Stephen Heath, Trans.). In *Image, music, text* (pp. 32–51). London: Fontana.

Beauchamp, Mark R., Bray, Steven R., & Albinson, John G. (2002). Pre-competition imagery, self-efficacy and performance in collegiate golfers. *Journal of Sports Sciences, 20*(9), 697–705.

Bickford, Maggie. (2005). The symbolic seasonal round in house and palace: Counting the auspicious nines in traditional China. In Ronald G. Knapp & Kai-Yin Lo (Eds.), *House, Home, Family: Living and Being Chinese* (pp. 348–371). Honolulu, Hawaii: University of Hawaii Press.

Boden, Margaret A. (2009). Computer models of creativity. *AI Magazine, Fall 2009.*

Bogost, Ian. (2007). *Persuasive Games: The Expressive Power of Videogames.* Cambridge, MA: MIT Press.

Bogost, Ian. (2013). Exploitationware. In Richard Colby, Matthew S. S. Johnson, & Rebekah Shultz Colby (Eds.), *Rhetoric/Composition/Play through Video Games: Reshaping Theory and Practice of Writing* (pp. 139–147). New York: Palgrave Macmillan US.

Bogost, Ian. (2015, March 13). Video games are better without characters. *The Atlantic.*

Booker, Christopher. (2004). *The Seven Basic Plots: Why We Tell Stories.* London; New York: Continuum.

Boyd, Brian. (2009). *On the Origin of Stories: Evolution, Cognition, and Fiction.* Cambridge, MA: Belknap Press of Harvard University Press.

Burke, Kenneth. (1969). *A Rhetoric of Motives.* Berkeley: University of California Press.

Campbell, Joseph. (2004). *The Hero with a Thousand Faces / Joseph Campbell* (Commemorative ed.). Princeton, NJ: Princeton University Press.

Chaiken, Shelly, & Trope, Yaacov (Eds.). (1999). *Dual-Process Theories in Social Psychology.* New York: Guilford Press.

Chan, Carina K. Y., & Cameron, Linda D. (2012). Promoting physical activity with goal-oriented mental imagery: A randomized controlled trial. *Journal of Behavioral Medicine, 35,* 347–363.

Chatman, Seymour Benjamin. (1978). *Story and Discourse: Narrative Structure in Fiction and Film.* Ithaca, NY: Cornell University Press.

Chow, Kenny K. N. (2021a). Crafting animated parables: An embodied approach to representing lifestyle behaviors for reflection. *Digital Creativity, 32*(1), 1–21.

Chow, Kenny K. N. (2021b). The influence of repeated interactions on the persuasiveness of simulation. *Interaction Studies, 22*(3), 373–395.

Chow, Kenny K. N. (2023). Simulation in the post-reality feedback loop. In Grant Hamilton & Carolyn Lau (Eds.), *Mapping the Posthuman* (pp. 61–76). New York: Routledge.

Cialdini, Robert B. (2009). *Influence: The Psychology of Persuasion* (EPub edition ed.). New York: HarperCollins Publishers.

Clear, James. (2018). *Atomic Habits: Tiny Changes, Remarkable Results: An Easy & Proven Way to Build Good Habits & Break Bad Ones.* New York: Avery.

Conroy, Dominic, Sparks, Paul, & Visser, Richard de. (2015). Efficacy of a non-drinking mental simulation intervention for reducing student alcohol consumption. *British Journal of Health Psychology, 20,* 688–707.

Coulson, Seana. (2001). *Semantic Leaps: Frame-Shifting and Conceptual Blending in Meaning Construction.* Cambridge: Cambridge University Press.

Davis, Fred D. (1989). Perceived usefulness, perceived ease of use, and user acceptance of information technology. *MIS Quarterly, 13*(3), 319–340.

Deci, Edward L., Betley, Gregory, Kahle, James, Abrams, Linda, & Porac, Joseph. (1981). When trying to win: Competition and intrinsic motivation. *Personality and Social Psychology Bulletin, 7*(1), 79–83.

Deterding, Sebastian, Dixon, Dan, Khaled, Rilla, & Nacke, Lennart. (2011). *From Game Design Elements to Gamefulness: Defining "Gamification".* Paper presented at the Proceedings of the 15th International Academic MindTrek Conference: Envisioning Future Media Environments.

Di Raimondo, Domenico, Tuttolomondo, A., Buttà, C., Casuccio, A., Giarrusso, L., Miceli, G., . . . Pinto, A. (2013). Metabolic and anti-inflammatory effects of a home-based programme of aerobic physical exercise. *International Journal of Clinical Practice, 67*(12), 1247–1253.

Duggan, William. (2015). *The Seventh Sense: How Flashes of Insight Change Your Life.* New York: Columbia University Press.

Escalas, Jennifer Edson, & Luce, Mary Frances. (2004). Understanding the effects of process-focused versus outcome-focused thought in response to advertising. *Journal of Consumer Research, 31*(2), 274–285.

Fauconnier, Gilles, & Turner, Mark. (2002). *The Way We Think: Conceptual Blending and the Mind's Hidden Complexities.* New York: Basic Books.

Feldman, Jerome. (2008). *From Molecule to Metaphor: A Neural Theory of Language.* Cambridge, MA: The MIT Press.

Fillmore, Charles J. (1985). Frames and the semantics of understanding. *Quaderni di Semantica, VI*(2), 222–254.

Fogg, B. J. (2003). *Persuasive Technology: Using Computers to Change What We Think and Do.* San Francisco, CA: Morgan Kaufmann Publishers.

Fogg, B. J. (2009). *A Behavior Model for Persuasive Design.* Paper presented at the Persuasive'09, Claremont, CA.

Forster, E. M. (1993). *Aspects of the Novel.* London: Hodder & Stoughton in association with Edward Arnold.

Freytag, Gustav. (1900). *Technique of the Drama: An Exposition of Dramatic Composition and Art* (Elias J. MacEwan, Trans. An authorized translation from the 6th German ed.). Chicago, IL: Scott, Foresman and Company.

Fromm, Erich. (1984). *The Fear of Freedom.* London: Ark.

Gaver, William W., Bowers, John, Boucher, Andrew, Gellerson, Hans, Pennington, Sarah, Schmidt, Albrecht, . . . Walker, Brendan. (2004). *The Drift Table: Designing for Ludic Engagement.* Paper presented at the Proceedings of CHI'04, Vienna, Austria.

Genette, Gérard. (1980). *Narrative Discourse: An Essay in Method.* Ithaca, NY: Cornell University Press.

Gentner, Dedre. (1983). Structure-mapping: A theoretical framework for analogy. *Cognitive Science, 7*, 155–170.

Gentner, Dedre, & Markman, Arthur B. (1997). Structure mapping in analogy and similarity. *American Psychologist, 52*(1), 45–56.

Gerrig, Richard J. (1993). *Experiencing Narrative Worlds: On the Psychological Activities of Reading.* New Haven, CT: Yale University Press.

González-Cutre, David, Sicilia, Álvaro, Sierra, Ana C., Ferriz, Roberto, & Hagger, Martin S. (2016). Understanding the need for novelty from the perspective of self-determination theory. *Personality and Individual Differences, 102*, 159–169.

González-Cutre, David, Romero-Elías, María, Jiménez-Loaisa, Alejandro, Beltrán-Carrillo, Vicente J., & Hagger, Martin S. (2020). Testing the need for novelty as a candidate need in basic psychological needs theory. *Motivation and Emotion, 44*(2), 295–314.

Graham, Benjamin. (1965). *The Intelligent Investor: A Book of Practical Counsel / Benjamin Graham* (3rd rev. ed.). New York: Harper & Row.

Grau, Oliver. (2003). *Virtual Art: From Illusion to Immersion.* Cambridge, MA: MIT Press.

Green, Melanie C. (2004). Transportation into narrative worlds: The role of prior knowledge and perceived realism. *Discourse Processes, 38*(2), 247–266.

Green, Melanie C., & Brock, Timothy C. (2000). The role of transportation in the persuasiveness of public narratives. *Journal of Personality and Social Psychology, 79*(5), 701–721.

Gregory, W. Larry, Cialdini, Robert B., & Carpenter, Kathleen M. (1982). Self-relevant scenarios as mediators of likelihood estimates and compliance: Does imagining make it so? *Journal of Personality and Social Psychology, 43*(1), 89–99.

Griffith, David A., & Rubera, Gaia. (2014). A cross-cultural investigation of new product strategies for technological and design innovations. *Journal of International Marketing, 22*(1), 5–20.

Grimm, Jacob, & Grimm, Wilhelm. (1884). *Grimm's Household Tales: With the Author's Notes* (Margaret Hunt, Trans.). London: George Bell and Sons.

Groom, Gloria, & Jill, Shaw. (2014). Cats. 27–33. Stacks of Wheat, 1890/91: Curatorial Entry. In Art Institute of Chicago (Ed.), *Monet Paintings and Drawings at the Art Institute of Chicago*. https://publications.artic.edu/monet/reader/paintingsanddrawings/section/135598

Guarnaccia, Steven. (2010). *The Three Little Pigs: An Architectural Tale*: Abrams Books for Young Readers.

Hall, Craig R., Mack, D. E., Paivio, Allan, & Hausenblas, H. A. (1998). Imagery use by athletes: Development of the sport imagery questionnaire. *International Journal of Sport Psychology, 29*(1), 73–89.

Hayles, N. Katherine. (1999). *How We Became Posthuman: Virtual Bodies in Cybernetics, Literature, and Informatics*. Chicago, IL: University of Chicago Press.

Hayles, N. Katherine. (2023). Posthuman bodies: Why they (still) matter. In Grant Hamilton & Carolyn Lau (Eds.), *Mapping the Posthuman* (pp. 29–48). New York: Routledge.

Hekkert, Paul, & Dijk, Matthijs van. (2011). *Vision in Design: A Guidebook for Innovators*. Amsterdam: BIS Publishers.

Herman, David, Jahn, Manfred, & Ryan, Marie-Laure (Eds.). (2005). *Routledge Encyclopedia of Narrative Theory*. London: Routledge.

Hershfield, Hal E., Goldstein, Daniel G., Sharpe, William F., Fox, Jesse, Yeykelis, Leo, Carstensen, Laura L., & Bailenson, Jeremy N. (2011). Increasing saving behavior through age-progressed renderings of the future self. *Journal of Marketing Research, 48*(SPL), S23–S37.

Hobeika, Lucie, Diard-Detoeuf, Capucine, Garcin, Béatrice, Levy, Richard, & Volle, Emmanuelle. (2016). General and specialized brain correlates for analogical reasoning: A meta-analysis of functional imaging studies. *Human Brain Mapping, 37*(5), 1953–1969.

Hoffman, Donald D. (1998). *Visual Intelligence: How We Create What We See*. New York: W.W. Norton.

Holland, Norman N. (2008). Spider-Man? Sure! The neuroscience of suspending dislief. *Interdisciplinary Science Reviews, 33*(4), 312–320.

Jacobs, Joseph. (1895). *English Fairy Tales*. New York: Grosset & Dunlap.

Johnson, Mark. (1987). *The Body in the Mind: The Bodily Basis of Meaning, Imagination, and Reason*. Chicago: University of Chicago Press.

Juul, Jesper. (2005). *Half-Real: Video Games between Real Rules and Fictional Worlds / Jesper Juul*. Cambridge, MA: The MIT Press.

Kelleter, Frank. (2017). Five ways of looking at popular seriality. In Frank Kelleter (Ed.), *Media of Serial Narrative* (pp. 7–34). Columbus: The Ohio State University Press.

Kim, Bang Hyun, Newton, Roberta A., Sachs, Michael L., Glutting, Joseph J., & Glanz, Karen. (2012). Effect of guided relaxation and imagery on falls self-efficacy: A randomized controlled trial. *Journal of the American Geriatrics Society (JAGS), 60*(6), 1109–1114.

King, Dominic, Greaves, Felix, Exeter, Christopher, & Darzi, Ara. (2013). 'Gamification': Influencing health behaviours with games. *Journal of the Royal Society of Medicine, 106*(3), 76–78.

Knäuper, Bärbel, McCollam, Amanda, Rosen-Brown, Ariel, Lacaille, Julien, Kelso, Evan, & Roseman, Michelle. (2011). Fruitful plans: Adding targeted mental imagery to implementation intentions increases fruit consumption. *Psychology and Health, 26*(5), 601–617.

Krippendorff, Klaus. (2006). *The Semantic Turn: A New Foundation for Design*. Boca Raton, FL: CRC/Taylor & Francis.

Lakoff, George. (1987). *Women, Fire, and Dangerous Things: What Categories Reveal about the Mind*. Chicago, IL: University of Chicago Press.

Lakoff, George. (2012). Explaining embodied cognition results. *Topics in Cognitive Science, 4*(4), 773–785.

Lakoff, George, & Johnson, Mark. (1999). *Philosophy in the Flesh: The Embodied Mind and Its Challenge to Western Thought*. New York: Basic Books.

Lakoff, George, & Johnson, Mark. (2003). *Metaphors We Live by*. Chicago, IL: University of Chicago Press.

Land, Norman E. (2013). Michelangelo and the stonecutters. *Source: Notes in the History of Art, 33*(1), 16–20.

Lang, Andrew. (1903). *The Crimson Fairy Book*. Retrieved from https://www.gutenberg.org/ebooks/2435

Lanier, Jaron. (2017). *Dawn of the New Everything: Encounters with Reality and Virtual Reality* (1st ed.). New York: Henry Holt and Company.

Ludden, Geke D.S., Rompay, Thomas J.L. van, Kelders, Saskia M., & Gemert-Pijnen, Julia E.W.C. van. (2015). How to increase reach and adherence of web-based interventions: A design research viewpoint. *Journal of Medical Internet Research, 17*(7).

Mandler, Jean M. (1992). How to build a baby: II. Conceptual primitives. *Psychological Review, 99*(4), 587–604.

Manzini, Ezio. (2015). *Design, When Everybody Designs: An Introduction to Design for Social Innovation*. Cambridge, MA: The MIT Press.

Margarita, Albedil. (2011). *Russian Lacquer Miniatures: Palekh, Mstiore, Fedoskino, Kholui* (Yury Pamfilov, Trans. Natalia Morozova Ed.): Jarkij gorod.

Markman, Arthur B., & Gentner, Dedre. (1993). Splitting the differences: A structural alignment view of similarity. *Journal of Memory and Language, 32*(4), 517–535.

Martin, Bella, & Hanington, Bruce. (2012). *Universal Methods of Design: 100 Ways to Research Complex Problems, Develop Innovative Ideas, and Design Effective Solutions.* Beverly, MA: Rockport Publishers.

Martin, Kathleen A., Moritz, Sandra E., & Hall, Craig R. (1999). Imagery use in sport: A literature review and applied model. *The Sport Psychologist, 13*(3), 245–268.

McKinley, William, Latham, Scott, & Braun, Michael. (2014). Organizational decline and innovation: Turnarounds and downward spirals. *The Academy of Management Review, 39*(1), 88–110.

Mekler, Elisa D., Brühlmann, Florian, Tuch, Alexandre N., & Opwis, Klaus. (2017). Towards understanding the effects of individual gamification elements on intrinsic motivation and performance. *Computers in Human Behavior, 71*, 525–534.

Mitchell, W. J. Thomas. (1986). *Iconology: Image, Text, Ideology*. Chicago, IL: University of Chicago Press.

Mollerup, Per. (2019). *Pretense Design: Surface over Substance*. Cambridge, MA: The MIT Press.

Nicholson, Scott. (2015). A RECIPE for meaningful gamification. In Torsten Reiners & Lincoln C. Wood (Eds.), *Gamification in Education and Business* (pp. 1–20). Cham: Springer International Publishing.

Norman, Donald A. (2002). *The Design of Everyday Things*. New York: Basic Books Inc.

Oh, Soo Youn, Bailenson, Jeremy, Weisz, Erika, & Zaki, Jamil. (2016). Virtually old: Embodied perspective taking and the reduction of ageism under threat. *Computers in Human Behavior, 60*, 398–410.

Oinas-Kukkonen, Harri. (2013). A foundation for the study of behavior change support systems. *Personal and Ubiquitous Computing, 17*(5), 1223–1235.

O'Sullivan, Sean. (2019). Six elements of serial narrative. *Narrative, 27*(1), 49–64.

Petty, Richard E., & Cacioppo, John T. (1986). The elaboration likelihood model of persuasion. *Advances in Experimental Social Psychology, 19*, 123–205.

Pfefferman, Richard. (2013). *Strategic Reinvention in Popular Culture: The Encore Impulse*. Basingstoke: Palgrave Macmillan.

Pham, Lien B., & Taylor, Shelley E. (1999). From thought to action: Effects of process-versus outcome-based mental simulations on performance. *Personality and Social Psychology Bulletin, 25*(2), 250–260.

Propp, Vladimir. (1968). *Morphology of the Folktale* (Laurence Scott, Trans. Louis A. Wagner Ed. 2nd ed.). Austin: University of Texas Press.

Propp, Vladimir. (2012). *The Russian Folktale* (Sibelan Forrester, Trans. Sibelan Forrester Ed.): Wayne State University Press.

Ramage, John D., Bean, John C., & Johnson, June. (2016). *Writing Arguments : A Rhetoric with Readings* (10th ed.). Boston, MA: Pearson.

Rapp, Amon. (2017). Drawing inspiration from world of warcraft: Gamification design elements for behavior change technologies. *Interacting with Computers, 29*(5), 648–678.

Rapp, Amon, & Cena, Federica. (2016). Personal informatics for everyday life: How users without prior self-tracking experience engage with personal data. *International Journal of Human-Computer Studies, 94*, 1–17.

Reiners, Torsten, & Wood, Lincoln C. (2015). *Gamification in Education and Business*. Cham: Springer.

Ritchie, L. David. (2017). *Metaphorical Stories in Discourse*. Cambridge: Cambridge University Press.

Rompay, Thomas J. L. van. (2008). Product expression: Bridging the gap between the symbolic and the concrete. In H. Schifferstein & Paul Hekkert (Eds.), *Product Experience* (pp. 333–351). San Diego, CA: Elsevier Science.

Ross, Elinor P. (April, 1979). Comparison of folk tale variants. *Language Arts, 56*(4), 422–426.

Rozhkovskaya, Natasha, & Reb, Michael. (2015). Is the list of incomplete open cubes complete? *Nexus Network Journal, 17*(3), 913–925.

Rubin, Gretchen. (2015). *Better Than Before: Mastering the Habits of Our Everyday Lives*. New York: Crown.

Ryan, Marie-Laure. (1990). Stacks, frames and boundaries, or narrative as computer language. *Poetics Today, 11*(4), 873–899.

Ryan, Richard M., & Deci, Edward L. (2000a). Intrinsic and extrinsic motivations: Classic definitions and new directions. *Contemporary Educational Psychology, 25*(1), 54–67.

Ryan, Richard M., & Deci, Edward L. (2000b). Self-determination theory and the facilitation of intrinsic motivation, social development, and well-being. *American Psychologist, 55*(1), 68–78.

Ryan, William S., Cornick, Jessica, Blascovich, Jim, & Bailenson, Jeremy N. (2019). Virtual reality: Whence, how and what for. In Albert "Skip" Rizzo & Stéphane Bouchard (Eds.),

Virtual Reality for Psychological and Neurocognitive Interventions (pp. 15–46). New York: Springer.

Salen, Katie, & Zimmerman, Eric. (2004). *Rules of Play: Game Design Fundamentals.* Cambridge, MA: The MIT Press.

Salter, Rebecca. (2006). *Japanese Popular Prints: From Votive Slips to Playing Cards.* University of Hawaii Press.

Sawyer, R. Keith. (2012). *Explaining Creativity: The Science of Human Innovation.* New York: Oxford University Press.

Seligman, Martin E. P., & Csikszentmihalyi, Mihaly. (2000). Positive psychology: An introduction. *American Psychologist, 55*(1), 5–14.

Sharp, John, & Macklin, Colleen. (2019). *Iterate: Ten Lessons in Design and Failure.* Cambridge, MA: The MIT Press.

Sherman, Steven J., Cialdini, Robert B., Schwartzman, Donna F., & Reynolds, Kim D. (1985). Imagining can heighten or lower the perceived likelihood of contracting a disease: The mediating effect of ease of imagery. *Personality and Social Psychology Bulletin, 11*(1), 118–127.

Shrout, Patrick E., & Fleiss, Joseph L. (1979). Intraclass correlations: Uses in assessing rater reliability. *Psychological Bulletin, 86*(2), 420–428.

Singer, Jerome L. (1999). Imagination. In Mark Runco & Steven Pritzker (Eds.), *Encyclopedia of Creativity* (1st ed., pp. 637–643). San Diego, CA: Academic Press.

Slowik, Mary. (2014). Telling 'What Is': Frame narrative in Zbig Rybczynski's Tango, Wendy Tilby and Amanda Forbis's When the Day Breaks, and Yuri Norstein's Tale of Tales. *Animation, 9*(3), 281–298.

Stets, Jan E., & Burke, Peter J. (2000). Identity theory and social identity theory. *Social Psychology Quarterly, 63*(3), 224–237.

Sutton-Smith, Brian. (2009). *The Ambiguity of Play* (Rev. ed.). Cambridge, MA: Harvard University Press.

Taylor, M. (2011). Imagination. In Mark Runco & Steven Pritzker (Eds.), *Encyclopedia of Creativity* (2nd ed., pp. 637–643). London: Academic Press.

Taylor, Shelley E., Pham, Lien B., Rivkin, Inna D., & Armor, David A. (1998). Harnessing the imagination: Mental simulation, self-regulation, and coping. *American Psychologist, 53*(4), 429–439.

Verganti, Roberto. (2016). *Overcrowded: Designing Meaningful Products in a World Awash with Ideas.* Cambridge, MA: The MIT Press.

Verplanken, Bas, & Aarts, Henk. (1999). Habit, attitude, and planned behaviour: Is habit an empty construct or an interesting case of goal-directed automaticity? *European Review of Social Psychology, 10*, 101–134.

Vygotsky, L. S. (1991). Imagination and creativity in the adolescent. *Soviet Psychology, 29*(1), 73–88.

Wardrip-Fruin, Noah, & Harrigan, Pat (Eds.). (2004). *First Person: New Media as Story, Performance, and Game.* Cambridge, MA: The MIT Press.

Western, Linda E. (April, 1980). A comparative study of literature through folk tale variants. *Language Arts, 57*(4), 395–402, 439.

Whitaker, Kirstie J., Vendetti, Michael S., Wendelken, Carter, & Bunge, Silvia A. (2018). Neuroscientific insights into the development of analogical reasoning. *Developmental Science, 21*(2), e12531.

Wiener, Norbert. (1961). *Cybernetics: Or, Control and Communication in the Animal and the Machine* (2nd ed.). New York: MIT Press.

Wood, Wendy. (2019). *Good Habits, Bad Habits: The Science of Making Positive Changes That Stick.* New York: Farrar, Straus and Giroux.

Worth, Keilah A., Sullivan, Helen W., Hertel, Andrew W., Rothman, Alexander J., & Jeffery, Robert W. (2005). Avoidance goals can be beneficial: A look at smoking cessation. *Basic and Applied Social Psychology, 27*(2), 107–116.

Wynd, Christine A. (2005). Guided health imagery for smoking cessation and long-term abstinence. *Journal of Nursing Scholarship, 37*(3), 245–250.

Yee, Nick, & Bailenson, Jeremy. (2006). *Walk a mile in digital shoes: The impact of embodied perspective-taking on the reduction of negative stereotyping in immersive virtual environments*. Paper presented at the PRESENCE 2006 conference: The 9th annual international workshop on presence, Cleveland, OH.

Yee, Nick, & Bailenson, Jeremy. (2007). The proteus effect: The effect of transformed self-representation on behavior. *Human Communication Research, 33*, 271–290.

Zeigarnik, Bluma. (2000). On finished and unfinished tasks. In Willis Davis Ellis (Ed.), *A Source Book of Gestalt Psychology* (pp. 300–314). London: Routledge.

傅連仲. (1982). 九九消寒圖. 紫禁城 *Forbidden City Magazine*, 31–32.

李文君. (2006). 一陽始新九九消寒. 紫禁城 *Forbidden City Magazine*, 6–9.

海上. (2005). 中國人的歲時文化: 嶽麓書社

潘力. (2022). 浮世繪*300*年， 極樂美學全覽: 任性出版.

胡櫨文. (2016). 消寒開泰—冬至圖象的使用脈絡. 故宮文物月刊, 395.

INDEX

Note: **Bold** page numbers refer to tables.